Energy Storage and Conversion

Development of Materials for Energy Storage and Conversion Devices

Energy Storage and Conversion

Development of Materials for Energy Storage and Conversion Devices

Editor

Masashi Kotobuki

Ming Chi University of Technology, Taiwan

NEW JERSEY · LONDON · SINGAPORE · BEIJING · SHANGHAI · HONG KONG · TAIPEI · CHENNAI

Published by

World Scientific Publishing Co. Pte. Ltd.
5 Toh Tuck Link, Singapore 596224
USA office: 27 Warren Street, Suite 401-402, Hackensack, NJ 07601
UK office: 57 Shelton Street, Covent Garden, London WC2H 9HE

British Library Cataloguing-in-Publication Data
A catalogue record for this book is available from the British Library.

ENERGY STORAGE AND CONVERSION
Development of Materials for Energy Storage and Conversion Devices

Copyright © 2025 by World Scientific Publishing Co. Pte. Ltd.

All rights reserved. This book, or parts thereof, may not be reproduced in any form or by any means, electronic or mechanical, including photocopying, recording or any information storage and retrieval system now known or to be invented, without written permission from the publisher.

For photocopying of material in this volume, please pay a copying fee through the Copyright Clearance Center, Inc., 222 Rosewood Drive, Danvers, MA 01923, USA. In this case permission to photocopy is not required from the publisher.

ISBN 978-981-98-0535-8 (hardcover)
ISBN 978-981-98-0536-5 (ebook for institutions)
ISBN 978-981-98-0537-2 (ebook for individuals)

For any available supplementary material, please visit
https://www.worldscientific.com/worldscibooks/10.1142/14102#t=suppl

Desk Editor: Rhaimie B Wahap

Typeset by Stallion Press
Email: enquiries@stallionpress.com

Printed in Singapore

Contents

Preface ix

Chapter 1: The effect of Y, Er co-doped on the sintering and electrical properties of $Mo_{0.05}Bi_{1.95}O_3$ electrolyte materials for solid oxide fuel cells   1
Weixing Qian, Hao Liang, Xuhang Zhu and Jihai Cheng

Chapter 2: A review on advanced optimization strategies of separators for aqueous zinc-ion batteries   19
Fukai Du, Fangfang Wu, Lu Ma, Jinxiu Feng, Xinyu Yin, Yuxi Wang, Xiaojing Dai, Wenxian Liu, Wenhui Shi and Xiehong Cao

Chapter 3: Effect of sintering temperature on phase transformation and energy storage properties of $0.95NaNbO_3$–$0.05Bi(Zn_{0.5}Zr^{0.5})O_3$ ceramics   53
Ying Ge, Dong Liu, Haifeng Zhang, Shuhao Yan, Bo Shi and Junjie Hao

Chapter 4: Heteroatom (N, P)-driven carbon nanomaterials for high-energy storage in supercapacitors   63
Yong Liu, Xian Zhang, Ke Zhang, Zifeng Wang and Guang Li,

vi Contents

Chapter 5: Impact of annealing on charge storage capability of thermally evaporated molybdenum oxide thin films   77
Sudesh Kumari, Rameez Ahmad Mir, Sanjay Upadhyay, O. P. Pandey and Anup Thakur

Chapter 6: V_2C-based lithium batteries: The influence of magnetic phase and Hubbard interaction   93
Jhon W. González, Sanber Vizcaya and Eric Suárez Morell

Chapter 7: Facile graphene quantum dot-anchoring strategy synthesis of single-atom iron–nitrogen electrocatalyst with enhanced ORR performance   109
Huinian Zhang, Suping Jia, Ning Li, Xiaolin Shi and Ziyuan Li

Chapter 8: Effect size of carbon micro-nanoparticles on cyclic stability and thermal performance of $Na_2SO4 \cdot 10H_2O$–$Na_2HPO_4 \cdot 12H_2O$ phase change materials   135
Zengbao Sun, Xin Liu, Shengnian Tie and Changan Wang

Chapter 9: An efficient $CoSe_2$-$Co3O_4$-Ag hybrid catalyst for electrocatalytic oxygen evolution   157
Qichen Liang, Nana Du and Huajie Xu

Chapter 10: Simulation and fabrication of titanium dioxide thin films for supercapacitor electrode applications   175
S. Harish, Muhammad Hamza, P. Uma Sathyakam, and Annamalai Senthil Kumar

Chapter 11: Ton-scale preparation of single-crystal Ni-rich ternary cathode materials for high-performance lithium ion batteries   193

Yongfu Cui, Jianzong Man, Leichao Meng, Wenjun Wang, Xueping Fan, Jing Yuan and Jianhong Peng

Chapter 12: Investigation on the influence of Zn content on the structural, optical, morphological and electrical properties of ternary compound $Cd_{1-x}Zn_xS$ window layer for CdTe solar cell   205

Ki Tong Hun, Kim Hyon Chol, and Jo Hye Gang

Index   227

Preface

The increasing global population and energy consumption have led to a rapid rise in the usage of fossil fuels in recent years. Consequently, there has been a significant increase in greenhouse gas emissions, which had detrimental effects on aquatic and animal life as well as the ecosystem. To address this problem, efficient energy storage and conversion devices with no greenhouse gas emissions, such as batteries, supercapacitors, and fuel cells, have been strongly required.

The development of suitable materials is essential for efficient energy storage and conversion devices. This special issue on "Energy storage and conversion" is organized to highlight significant breakthroughs in energy storage and conversion systems, focusing on material development.

The separator of batteries is a crucial component for long-term battery performance. In a brief review, Du et al. summarized the development history of separators and their surface modification for aqueous zinc-ion batteries.[1] Gonzalez et al. showed the effect of the magnetic phase on the characteristics of monolayer MXene V_2C used as an electrode for lithium-ion batteries.[2] Cui et al. demonstrated ton-scale production of single-crystal nickel (Ni)-rich cathode materials ($Li[Ni_xCo_yMn_{1-x-y}]O_2$) with an initial discharge capacity of 169 mAh/g at 0.1 C.[3] The impacts of carbon micro-nanoparticle sizes on the thermal properties and dispersion stability of $Na_2SO_4 \cdot 10H_2O$–$Na_2HPO_4 \cdot 12H_2O$ (EHS) phase transition materials were investigated by Sun et al.[4] Utilising 0.95$NaNbO_3$–0.05Bi$(Zn_{0.5}Zr_{0.5})O_3$ ceramics, Ge et al. were able to obtain the high recoverable energy storage density of 0.74 J/cm^3 and efficiency (η) of

71% (at 140 kV/cm) at 1150ΔC.[5] N and P doped nanoflower-like carbon sphere maintained 95.2% of its initial specific capacitance of 274.9 F/g after 4000 cycles.[6] The uniform TiO_2 nanotube film electrochemically anodized over a titanium foil had an aerial capacitance of 1.0193 F/cm^2 at 10 mV/s, whereas the etched one had a 12.8764 F/cm^2 at 10 mV/s using a 1 M HCl.[7] The capacitive behaviour of MoO_3 thin films (0.179 F/g) was demonstrated by Kumari et al., paving the way for their application in a variety of energy storage devices.[8]

Energy conversion systems like fuel cells, solar cells etc. are also important for the development of sustainable communities. $(YO_{1.5})_{0.20}(MoO_3)_{0.05}(ErO_{1.5})_{0.05}(BiO_{1.5})_{0.7}$ showed remarkable conductivity of 16×10^{-2} S cm^{-1} and high thermal stability for an electrolyte of solid oxide fuel cell by Qian et al.[9] The electrical resistivity of $Cd_{1-x}Zn_xS$ thin films in CdTe solar cells increase with Zn content and optimum in the range x = 0.5-0.6.[10] Liang et al. found that flower-like $CoSe_2$-Co_3O_4-Ag has a reasonably high electrocatalytic activity for OER, with an overpotential of 324 mV at a current density of 10 mA/cm^2.[11] There is a lot of potential for single-atom catalysts, particularly atomically distributed Fe–Nx–C, to replace Pt-based electrocatalysts in the ORR. In 0.1 M KOH environment, the Fe–N-GQDs/PC electrocatalyst demonstrates excellent electrocatalytic ORR activity, with half-wave potentials of 0.84 against 0.85 V for Pt/C.[12]

Eco-friendly, effective energy storage and conversion systems are essential for building a sustainable society. This special issue will make a substantial contribution to the development of efficient energy storage and conversion systems.

References

1. F. Du et al., Func. Mat. Letters **16**, 2340017 (2023).
2. J. W. Gonzalez et al., Func. Mat. Letters **16**, 2340023 (2023).
3. Y. Cui et al., Func. Mat. Letters **16**, 2340035 (2023).
4. Z. Sun et al., Func. Mat. Letters **16**, 2340032 (2023).
5. Y. Ge et al., Func. Mat. Letters **16**, 235001 (2023).

6. Y. Liu *et al.*, *Func. Mat. Letters* **16**, 2351010 (2023).
7. S. Harish *et al.*, *Func. Mat. Letters* **16**, 2340034 (2023).
8. S. Kumari *et al.*, *Func. Mat. Letters* **16**, 2340018 (2023).
9. W. Qian *et al.*, *Func. Mat. Letters* **16**, 2351007 (2023).
10. K. T. Hun, *Func. Mat. Letters* **16**, 2340038 (2023)
11. Q. Liang *et al.*, *Func. Mat. Letters* **16**, 2340033 (2023).
12. H. Zhang *et al.*, *Func. Mat. Letters* **16**, 2340031 (2023).

Masashi Kotobuki
Ming Chi University of Technology, Taiwan

Chapter 1

The effect of Y, Er co-doped on the sintering and electrical properties of $Mo_{0.05}Bi_{1.95}O_3$ electrolyte materials for solid oxide fuel cells

Weixing Qian, Hao Liang, Xuhang Zhu and Jihai Cheng*

School of Energy Materials and Chemical Engineering
Hefei University, Hefei 230022, P. R. China
*cjh@hfuu.edu.cn

Bismuth oxide-based electrolyte materials co-doping with yttrium and erbium were synthesized by nitrate combustion method. The crystal structure was analyzed by X-ray diffraction. The data on electrical conductivity were carried out using AC impedance spectroscopy. The results suggested that all of the co-doped bismuth oxide samples obtained δ-phase with cubic fluorite structure. The conductivity of $(YO_{1.5})_{0.2}(MoO_3)_{0.05}(ErO_{1.5})_{0.05}(BiO_{1.5})_{0.7}$ ceramic electrolyte sintered at 925°C for 10 h reached 12.26×10^{-2} S cm^{-1} at 700°C. It has good thermal stability and is a promising electrolyte material for SOFC.

Keywords: Solid oxide fuel cells; electrolyte; bismuth oxide; electrical properties.

1. Introduction

Energy saving and green living are the urgent needs of people today. In an era when the world is calling for low carbon and environmental protection, researchers have proposed numerous ideas for fuel cells which utilize clean energy.[1,7] Among them, the solid-state

---

*Corresponding authors.
To cite this article, please refer to its earlier version published in the Functional Materials Letters, Volume 16(8), 2351007 (2023), DOI: 10.1142/S1793604723510074.

structure of solid oxide fuel cell (SOFC) stands out as a safe and highly efficient device for converting electrical energy. The earliest developed and commonly utilized SOFC electrolyte material is Yttria Stabilized Zirconia (YSZ), however, it exhibits some drawbacks such as high working temperature (800–1000°C) and slow start-up time.[8,9] Hence, it is worthy to search for a solid electrolyte material that can achieve high conductivity and operate at low to medium temperature for its substitution.

Bismuth oxide is a polycrystalline material. Monoclinic structure of α-Bi_2O_3 is stable at room temperature to 730°C. Pure bismuth oxide undergoes a phase transition to the cubic fluorite phase (δ-Bi_2O_3) at 730°C to the melting point (825°C), which is a superior oxygen ion conductor due to its 25% oxygen vacancies and the polarization properties of bismuth ions, resulting in extremely high electrical conductivity.[10] Upon cooling at 650°C, δ-Bi_2O_3 transforms into a tetragonal structure (β-phase) and body-centered cubic structure (γ-phase), but neither of them shows adequate conductivity.[11] Doping with certain elements can stabilize δ-Bi_2O_3 to room temperature. İsmail Ermiş and Shaikh[12] reported that doped bismuth oxide was prepared by the solid state reaction method and δ-phase was obtained. The results showed that the long time high temperature heat treatment could increase the diffusion rate of Tb and Gd ions to Bi_2O_3. $(Bi_2O_3)_{0.90}(Tb_4O_7)_{0.05}(Gd_2O_3)_{0.05}$ was tested and was found to have the capability to obtain a good conductivity of 3.88×10^{-1} S cm^{-1} at 850°C. Ayhan Güldeste investigated Dy–Eu–Tm co-doped bismuth oxide system.[13] They found that the second calcination may stabilize the cubic phase after conductivity measurement. Cardenas-Terrazas studied dysprosium and tantalum-doped bismuth oxide electrolyte.[14] The $(Dy_2O_3)_{0.13}(Ta_2O_5)_{0.02}(Bi_2O_3)_{0.85}$ system showed the highest ionic conductivity at 8×10^{-2} S cm^{-1} at 500°C. The minimum activation energy was 0.20 eV.

Previous reports suggest that doping with tungsten can effectively suppress the transformation of δ-Bi_2O_3 in the cubic phase fluorite structure at medium and high temperatures to α-Bi_2O_3 in the monoclinic structure at room temperature, thereby maintaining the cubic phase fluorite structure of δ-Bi_2O_3 at room temperature.[15–19]

Due to the similar chemical properties of tungsten and molybdenum, doping with molybdenum has also been reported in other literatures as a means of stabilizing the δ phase.[20,22] In this study, we fixed the concentration of molybdenum at 5% and investigate the effects of different doping ratios of yttrium and erbium on the microstructure and electrochemical properties of doped bismuth oxide.

2. Materials and Methods

$(YO_{1.5})_x(MoO_3)_{0.05}(ErO_{1.5})_y(BiO_{1.5})_{0.95-x-y}$ (x = 0.05, 0.10, 0.15, 0.20, y = 0.05, 0.10, 0.15, 0.20) and an undoped bismuth oxide sample were prepared by nitrate combustion method, with the latter serving as a control experiment for structural transformation. $Y(NO_3)_3 \cdot 6H_2O$ (AR, 99.5%, Macklin), $Er(NO_3)_3 \cdot 5H_2O$ (AR, 99.9%, Macklin) and $(NH_4)_6Mo_7O_{24} \cdot 4H_2O$ (AR, 99%, Sinopharm Chemical Reagent) were dissolved in deionized water according to the designed stoichiometric ratio. Glycine (AR, 99.5%, Macklin) was added at 1.8 times the molar ratio of metal ions.[23] The solution was stirred at a constant temperature of 60°C until clarified. After that, 5 ml of concentrated nitric acid was added and a certain amount of bismuth nitrate $Bi(NO_3)_3 \cdot 5H_2O$ (AR, 99%, Macklin) was weighed and added to the beaker. After stirring the solution again until transparent, it was poured into an evaporating dish and heated on an electric stove until it underwent spontaneous combustion. The combusted and well ground precursor powder was calcined in a muffle furnace at 800°C for 15 h. The calcined powder was pressed into 14 mm diameter pellets and 35 mm × 4 mm × 4 mm rectangular strips respectively at 200 Mpa. The samples were allowed to sinter at different temperatures for 10 h depending on the composition of the elements. For example, the sample with $x + y = 0.1$ was sintered at 850°C. In order to prevent the electrolyte sheet from melting during sintering at high temperatures and to achieve high density, it is necessary to adjust the sintering temperature by increasing 25°C for every 5% decrease in bismuth concentration. All samples were calcined and sintered at a heating rate of 2°C min^{-1} then slowly cooled to room temperature at a rate of 5°C min^{-1}.

4 W. Qian et al.

The phase formation characteristics of the electrolyte pellets were determined by X-ray diffraction analysis. Cell parameters and cell volumes were calculated by Jade6 and Unitcellwin software. The electrical conductivity of the samples was tested at 400–800°C by an electrochemical workstation using the AC impedance method. Thermal expansion coefficient (TEC) of the rectangular sintered strips was measured from room temperature to 800°C using an expansion analyzer (DIL0809PC).

3. Results and Discussions

Figure 1 shows the XRD patterns of undoped bismuth oxide and $(YO_{1.5})_x(MoO_3)_{0.05}(ErO_{1.5})_y(BiO_{1.5})_{0.95-x-y}$ electrolyte pellets after sintering at different temperatures. For convenience, we have designated the samples as $Y_5Mo_5Er_5$, $Y_{10}Mo_5Er_5$, $Y_{15}Mo_5Er_5$, $Y_{20}Mo_5Er_5$, etc. according to their respective compositions. The results show that the undoped bismuth oxide sample exhibits a monoclinic structure consistent with PDF standard card No. 65-2366 upon cooling to room temperature, whereas the doped samples form face-centered cubic structures at room temperature after sintering at the corresponding temperature for 10 h (Table 1). Y^{3+}, Mo^{6+}, and Er^{3+} enter into the lattice of bismuth oxide because of doping, which are consistent with PDF standard card No. 52-7001 for δ-Bi_2O_3. Additionally, no other extraneous peaks exist, indicating that the doping process can stabilize the δ phase to room temperature.

As can be seen from Fig. 2, unit cell parameters decrease when doping amount grows up. This is because the proportion of Bi^{3+} in the sample gradually decreases, and the ionic radius of Y^{3+} (0.09 nm), Er^{3+} (0.089 nm), and Mo^{6+} (0.059 nm) are smaller than that of Bi^{3+} (0.103 nm), which leads to the shrinkage of the unit cell.[24] This is the reason why the unit cell parameter of sample $Y_5Mo_5Er_5$ is closest to pure δ-Bi_2O_3 (0.565 nm). Similar reports have been found in other literatures.[25,26]

The SEM image of $(YO_{1.5})_x(MoO_3)_{0.05}(ErO_{1.5})_y(BiO_{1.5})_{0.95-x-y}$ samples is shown in Fig. 3. As the doping amount of yttrium or erbium

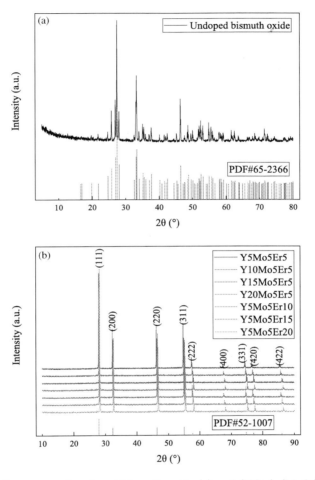

Fig. 1. XRD patterns of undoped bismuth oxide (a) and $(YO_{1.5})_x(MoO_3)_{0.05}(ErO_{1.5})_y(BiO_{1.5})_{0.95-x-y}$ samples (b).

increases, the grain size gradually decreases, which corresponds to the calculated decrease of the unit cell parameters. The doping of Y^{3+} presents a more significant effect on the grain size reduction than Er^{3+}. The sample $Y_{20}Mo_5Er_5$ exhibits the smallest grain size of 0.7–2 μm. Moreover, an increase in the sintering temperature leads to a decrease in the pores on the electrolyte surface, which indicates

Table 1. Sample code, unit cell parameters and other data of $(YO_{1.5})_x(MoO_3)_{0.05}(ErO_{1.5})_y(BiO_{1.5})_{0.95-x-y}$ system.

Sample	Composition	Sintering temperature	Phase	Unit cell parameters (nm)	Cell volume (nm^3)
$Y_5Mo_5Er_5$	$(YO_{1.5})_{0.05}(MoO_3)_{0.05}(ErO_{1.5})_{0.05}(BiO_{1.5})_{0.85}$	850°C	δ	0.5561	0.1719
$Y_{10}Mo_5Er_5$	$(YO_{1.5})_{0.1}(MoO_3)_{0.05}(ErO_{1.5})_{0.05}(BiO_{1.5})_{0.8}$	875°C	δ	0.5543	0.1703
$Y_{15}Mo_5Er_5$	$(YO_{1.5})_{0.15}(MoO_3)_{0.05}(ErO_{1.5})_{0.05}(BiO_{1.5})_{0.75}$	900°C	δ	0.5521	0.1683
$Y_{20}Mo_5Er_5$	$(YO_{1.5})_{0.2}(MoO_3)_{0.05}(ErO_{1.5})_{0.05}(BiO_{1.5})_{0.7}$	925°C	δ	0.5504	0.1668
$Y_5Mo_5Er_{10}$	$(YO_{1.5})_{0.05}(MoO_3)_{0.05}(ErO_{1.5})_{0.10}(BiO_{1.5})_{0.8}$	875°C	δ	0.5541	0.1701
$Y_5Mo_5Er_{15}$	$(YO_{1.5})_{0.05}(MoO_3)_{0.05}(ErO_{1.5})_{0.15}(BiO_{1.5})_{0.75}$	900°C	δ	0.5520	0.1682
$Y_5Mo_5Er_{20}$	$(YO_{1.5})_{0.05}(MoO_3)_{0.05}(ErO_{1.5})_{0.2}(BiO_{1.5})_{0.7}$	925°C	δ	0.5499	0.1663

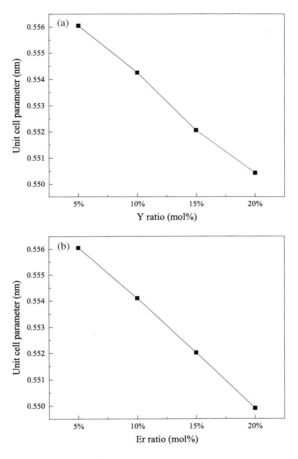

Fig. 2. The effects of Y (a) and Er (b) ratio on unit cell parameters.

a higher density of the electrolyte. The relative density Q can be measured using the Archimedes' drainage method, calculated by the following equation:

$$Q = \frac{M_1 - M_2}{M_3 - M_2} \times 100\%, \tag{1}$$

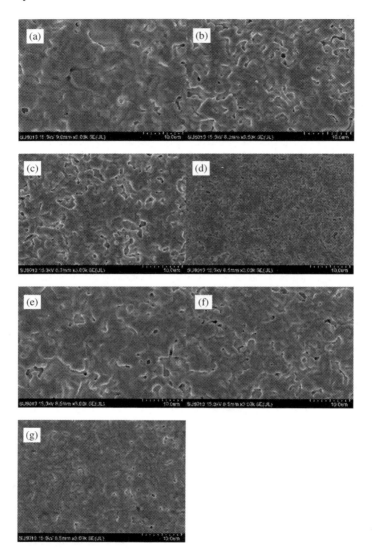

Fig. 3. The surface SEM micrographs of $(YO_{1.5})_x(MoO_3)_{0.05}(ErO_{1.5})_y(BiO_{1.5})_{0.95-x-y}$ samples.

where M_1 is the weight of the dried ceramic pellet. It is submerged in the boiling water for the purpose of removing the air inside. Then, the weight of the ceramic pellet that is suspended in the water is called M_2. The ceramic pellet is taken out, wiped off the water on the surface, at which time the weight is called M_3. The relative density of sample $Y_{20}Mo_5Er_5$ is calculated to be 95.39%, which meets the density requirement of SOFC electrolyte.

Figure 4 shows the energy dispersive spectroscopy (EDS) of sample $Y_5Mo_5Er_5$. The weight and atoms percentages of each element are shown in Table 2. The data show that the contents of Y, Er, and Mo are higher than those of chemical formula $(YO_{1.5})_{0.05}$

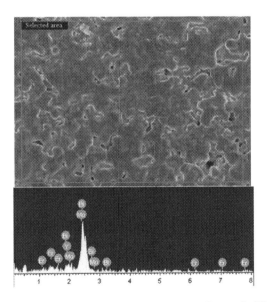

Fig. 4. The energy dispersive spectroscopy of sample $Y_5Mo_5Er_5$.

Table 2. Weight and atoms percentages of each element of sample $Y_5Mo_5Er_5$.

Element	Weight percentage (%)	Atoms percentage (%)
Y	3.61	7.58
Mo	4.33	8.42
Er	7.78	8.68
Bi	84.29	75.32

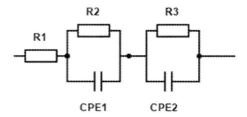

Fig. 5. The equivalent circuit for testing.

$(MoO_3)_{0.05}(ErO_{1.5})_{0.05}(BiO_{1.5})_{0.85}$. This is because the sintering temperature of sample $Y_5Mo_5Er_5$ exceeded the melting point of bismuth oxide, resulting in a decrease of Bismuth element content on the surface while the other elements increased in equal proportion.

The impedance of the samples were tested using $R_1(QR_2)(QR_3)$ as an equivalent circuit (Fig. 5), where R_1 represents the grain resistance, R_2 is the grain boundary resistance, R_3 is the electrode contact resistance, and Q is the constant phase angle element (CPE). The total resistance is defined as follows:

$$R = R_1 + R_2. \qquad (2)$$

Figure 6 shows the impedance spectrums of sample $Y_{20}Mo_5Er_5$ at different temperatures. The EIS spectra of SOFC electrolyte ideally comprise three semicircular arcs. The high-frequency section corresponds to the grain impedance (R_1), the mid-frequency region corresponds to the grain boundary resistance (R_2), and the low-frequency range corresponds to the interfacial resistance between the electrode and the electrolyte (R_3). Due to the long silver wire connecting both sides of the electrolyte sheet and the low resistance of bismuth oxide at high temperatures, ultra-high frequency inductance occurs within this system, so the part of the inductance below the solid axis is truncated in the figure. In addition, the small time constant of the grain impedance and the limitations of the electrochemical workstation's performance, no high frequency semicircle corresponding to the grain impedance is observed.[27] At 400°C, the

The effect of Y, Er co-doped on the sintering and electrical properties 11

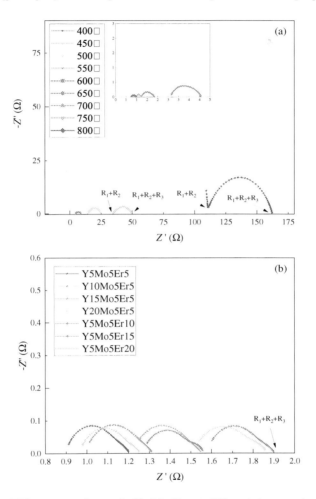

Fig. 6. The EIS spectra of sample $Y_{20}Mo_5Er_5$ at different temperatures (a) and all samples at 700°C (b).

semicircle arc indicates the grain boundary resistance (R_2), which becomes invisible at higher temperatures, leaving only the arc representing the impedance at the electrode interface. As the temperature increases, the intercept of the impedance arc with the real axis decreases, indicating that the summary of the grain resistance and the grain boundary resistance is decreasing.[28]

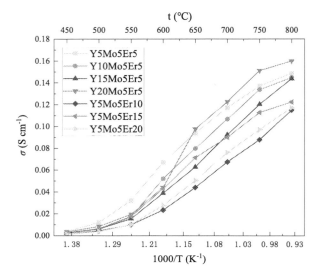

Fig. 7. The conductivity curves of different samples.

This decreasing trend is diminishing, which can also be seen clearly in Fig. 7. The conductivity of the sample $Y_{20}Mo_5Er_5$ shows a rapid increase from 600°C to about 650°C, followed by two subsequent stages in which the conductivity continues to increase with increasing temperature. However, the rate of increase gradually diminishes in these stages. At the beginning, the increased amount of Y^{3+} results in a decreasing conductivity performance. Nevertheless, At a Y^{3+} content of 20%, the conductivity of $Y_{20}Mo_5Er_5$ undergoes a remarkable transformation, reaching 12.62×10^{-2} S cm^{-1} at 700°C, which is comparable to that of other electrolyte materials, demonstrating the potential of $Y_{20}Mo_5Er_5$ as an electrolyte for SOFC. Table 3 presents the conductivities of various Bi_2O_3 electrolyte materials doped with different elements. The reason is that Y^{3+} replaces Bi^{3+}, resulting in a large number of oxygen vacancies, while the sample $Y_{20}Mo_5Er_5$ possesses a high dense density, which accelerates the migration rate of oxygen ions.

The Arrhenius curves in Fig. 8 reveal that the curves of all the samples are divided into two parts: Low temperature (LT) and high temperature (HT). All of them satisfy the Arrhenius relationship

Table 3. Conductivities of various Bi_2O_3 electrolyte materials doped with different elements.

Materials	Testing temperature (°C)	σ (S cm^{-1})	Reference
$(YO_{1.5})_{0.20}(MoO_3)_{0.05}(ErO_{1.5})_{0.05}(BiO_{1.5})_{0.7}$	700	12.62×10^{-2}	This work
$(Bi_2O_3)_{0.85}(Gd_2O_3)_{0.1}(Lu_2O_3)_{0.05}$	650	9.2×10^{-2}	29
$(Tb_4O_7)_{0.0125}(Gd_2O_3)_{0.025}(Ho_2O_3)_{0.05}(Bi_2O_3)_{0.9125}$	750	11.9×10^{-2}	25

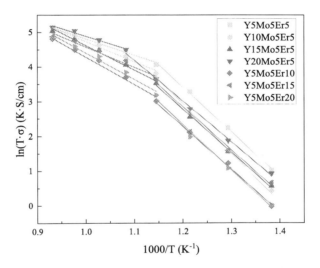

Fig. 8. The Arrhenius plots of conductivities from 450°C to 800°C.

equation. According to some literatures, these turning points are called the order–disorder transformation.[13,30] The curve for the rest of the samples turns at 600°C, while the sample $Y_{20}Mo_5Er_5$ once again shows its distinction: Its breakthrough occurs at 650°C, which is attributed to the increase in ion mobility. The disordered structure facilitates oxygen ions to find migration paths and increase the ionic conductivity, which becomes a factor of activation energy reduction.[31] The Arrhenius equation is expressed as follows:

$$\sigma = \frac{A}{T} \times e^{\left(\frac{Ea}{KT}\right)}, \tag{3}$$

where σ stands for electrical conductivity. A is the pre-exponential factor. T is the absolute temperature. K is Boltzmann's constant. The activation energy Ea can be obtained from the reverse derivation of Arrhenius formula. The activation energy represents the energy barrier that the carriers need to overcome to migrate. As can be seen from Table 4, the migration of oxygen ions is restricted at low temperatures, resulting in generally high activation energy. A sample $Y_{20}Mo_5Er_5$ presents the lowest activation energy in the high temperature range, reaching 0.349 eV. The reaction requires the least amount of energy and is the fastest and easiest to perform, which is another reason for the highest conductivity.

To prevent the cell fracture and reduce the electrical performance, it is essential to match the thermal expansion coefficients (TEC) of the SOFC electrolyte with cathode and anode materials. Figure 9 shows the thermal expansion curves of samples $Y_5Mo_5Er_5$, $Y_{10}Mo_5Er_5$, and $Y_{20}Mo_5Er_5$. As can be seen from the figure, the curve of sample $Y_{20}Mo_5Er_5$ is almost linear, while the remaining two samples change at about 610°C, which also proves that the sample

Table 4. The conductivity activation energy of different samples.

Sample code	Activation energy of LT (eV)	Activation energy of HT (eV)
$Y_5Mo_5Er_5$	1.0317	0.3755
$Y_{10}Mo_5Er_5$	1.1348	0.4660
$Y_{15}Mo_5Er_5$	1.3956	0.5707
$Y_{20}Mo_5Er_5$	0.9425	0.3490
$Y_5Mo_5Er_{10}$	0.9875	0.6353
$Y_5Mo_5Er_{15}$	0.9955	0.4716
$Y_5Mo_5Er_{20}$	1.0474	0.6134

The effect of Y, Er co-doped on the sintering and electrical properties 15

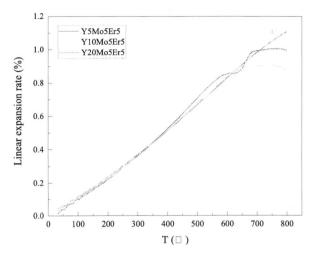

Fig. 9. The thermal expansion curves of sample $Y_5Mo_5Er_5$, $Y_{10}Mo_5Er_5$, and $Y_{20}Mo_5Er_5$.

has good thermal stability. We used the average thermal expansion coefficient, calculated as follows:

$$\alpha = \frac{1}{L_0} \times \frac{D_1 - D_0}{T_1 - T_0}, \qquad (4)$$

where T_0 and T_1 denote the starting temperature and the actual temperature at the time of testing, respectively. L_0 is the initial length of the sintered strip. D_0 represents the displacement of the sample at T_0. D_1 is the displacement of the sample at T_1. L_1 is the length of the sintered strip at T_1. The TEC of sample $Y_{20}Mo_5Er_5$ is calculated to be $13.21 \times 10^{-6}/K$, demonstrating its numerical compatibility with cathode materials such as $Sr_2Fe_{1.5}Mo_{0.5}O_6$ (SFM).[32]

4. Conclusions

The $(YO_{1.5})_x(MoO_3)_{0.05}(ErO_{1.5})_y(BiO_{1.5})_{0.95-x-y}$ (x = 0.05, 0.10, 0.15, 0.20, y = 0.05, 0.10, 0.15, 0.20) and pure Bi_2O_3 electrolyte materials were prepared by the nitrate combustion method. The results of

XRD demonstrate that doping enables the electrolyte materials to maintain a cubic fluorite structure at room temperature. Notably, $(YO_{1.5})_{0.20}(MoO_3)_{0.05}(ErO_{1.5})_{0.05}(BiO_{1.5})_{0.7}$ exhibits good thermal stability and exceptional conductivity, while also meeting the density requirements of electrolytes for SOFC. As the temperature increases, its conductivity increases, reaching 16×10^{-2} S cm^{-1} at 800°C. Hence, $(YO_{1.5})_{0.20}(MoO_3)_{0.05}(ErO_{1.5})_{0.05}(BiO_{1.5})_{0.7}$ is a promising electrolyte material for SOFC.

Acknowledgments

This project is supported by the Nature Science Foundation of Anhui Province of China (Grant No. 2108085ME152) and the Talent Research Fund Project of Hefei University (Grant No. 21-22RC34).

References

1. A. V. Shlyakhtina, N. V. Lyskov, I. V. Kolbanev, A. N. Shchegolikhin, O. K. Karyagina and L. G. Shcherbakova, *Int. J. Hydrog. Energy* **46**, 16989 (2021).
2. A. G. Jolley, R. Jayathilake and E. D. Wachsman, *Ionics* **25**, 3531 (2019).
3. M. Singh and A. K. Singh, *Int. J. Hydrog. Energy* **45**, 24014 (2020).
4. N. Momin, J. Manjanna, S. T. Aruna, S. Senthilkumar, D. S. Reddy and A. Kumar, *J. Chem. Sci.* **134**, 37 (2022).
5. M. Irshad, M. Khalid, M. Rafique, A. N. Tabish, A. Shakeel, K. Siraj, A. Ghaffar, R. Raza, M. Ahsan, Q. T. Ain and Q. U. Ain, *Sustainability* **13**, 12595 (2021).
6. X.-W. Zhou, Y.-F. Sun, G.-Y. Wang, T. Gao, K. T. Chuang, J.-L. Luo, M. Chen and V. I. Birss, *Electrochem. Commun.* **43**, 79 (2014).
7. P. Wang, C. Fu, W. Zhan, R.-H. Zhang, L.-Y. Yan, Z. Cheng, X. Zhang and X.-W. Zhou, *Ceram. Int.* **49**, 2319 (2023).
8. B. Singh, S. Ghosh, S. Aich and B. Roy, *J. Power Sources* **339**, 103 (2017).

9. Z. Zakaria, S. H. Abu Hassan, N. Shaari, A. Z. Yahaya and Y. Boon Kar, *Int. J. Energy Res.* **44**, 631 (2020).
10. M. Drache, P. Roussel and J.-P. Wignacourt, *Chem. Rev.* **107**, 80 (2007).
11. A. M. Azad, S. Larose and S. A. Akbar, *J. Mater. Sci.* **29**, 4135 (1994).
12. İ. Ermiş and S. P. S. Shaikh, *Ceram. Int.* **44**, 18776 (2018).
13. A. Güldeste, M. Aldoori, M. Balci, M. Ari and Y. Polat, *J. Rare Earths* **41**, 406 (2022).
14. P. S. Cardenas-Terrazas, M. T. Ayala-Ayala, J. Muñoz-Saldaña, A. F. Fuentes, D. A. Leal-Chavez, J. E. Ledezma-Sillas, C. Carreño-Gallardo and J. M. Herrera-Ramirez, *Ionics* **26**, 4579 (2020).
15. A. A. Lidie, K. T. Lee and E. D. Wachsman, *ECS Trans.* **50**, 15 (2013).
16. D. W. Jung, K. T. Lee and E. D. Wachsman, *J. Korean Ceram. Soc.* **51**, 260 (2014).
17. A. Borowska-Centkowska, M. Leszczynska, W. Wrobel, M. Malys, M. Krynski, S. Hull, F. Krok and I. Abrahams, *Solid State Ion.* **308**, 61 (2017).
18. A. Borowska-Centkowska, M. Leszczynska, W. Wrobel, M. Malys, S. Hull, F. Krok and I. Abrahams, *Solid State Ion.* **345**, 115173 (2020).
19. E. P. Kharitonova, E. I. Orlova, N. V. Gorshkov, V. G. Goffman and V. I. Voronkova, *Ceram. Int.* **47**, 31168 (2021).
20. E. I. Orlova, E. P. Kharitonova, N. V. Gorshkov, V. G. Goffman and V. I. Voronkova, *Solid State Ion.* **302**, 158 (2017).
21. E. P. Kharitonova, E. I. Orlova, N. V. Gorshkov, V. G. Goffman and V. I. Voronkova, *Ceram. Int.* **44**, 12886 (2018).
22. E. P. Kharitonova, E. I. Orlova, N. V. Gorshkov, V. G. Goffman, S. A. Chernyak and V. I. Voronkova, *J. Alloys Compd.* **787**, 452 (2019).
23. A. Raghvendra and P. Singh, *Ceram. Int.* **43**, 11692 (2017).
24. A. Dapčević, D. Poleti, J. Rogan, A. Radojković, M. Radović and G. Branković, *Solid State Ion.* **280**, 18 (2015).
25. M. Balci, A. Cengel and M. Ari, *Chin. J. Phys.* **79**, 89 (2022).
26. P. Guler, M. S. Duyar and S. Yilmaz, *Chem. Papers* **76**, 5513 (2022).
27. J. Cheng, R. Xu and Y. Shi, *J. Rare Earths* **39**, 728 (2021).
28. J. Yang, B. Ji, J. Si, Q. Zhang, Q. Yin, J. Xie and C. Tian, *Int. J. Hydrog. Energy* **41**, 15979 (2016).

29. Y. Polat, H. Akalan and M. Arı, *Int. J. Hydrog. Energy* **42**, 614 (2017).
30. T. B. Tran and A. Navrotsky, *Phys. Chem. Chem. Phys.* **16**, 2331 (2014).
31. S. Boyapati, E. D. Wachsman and N. Jiang, *Solid State Ion.* **140**, 149 (2001).
32. W. Pan, Q. Weixing, X. Ronghao and C. Jihai, *Process. Appl. Ceram.* **16**, 64 (2022).

Chapter 2

A review on advanced optimization strategies of separators for aqueous zinc-ion batteries

Fukai Du[*], Fangfang Wu[*,‡,¶], Lu Ma[*], Jinxiu Feng[*], Xinyu Yin[*], Yuxi Wang[*], Xiaojing Dai[*], Wenxian Liu[*], Wenhui Shi[†] and Xiehong Cao[*,§,¶]

[*]*College of Materials Science and Engineering, Zhejiang University of Technology, Hangzhou 310014, P. R. China*
[†]*Center for Membrane Separation and Water Science & Technology, College of Chemical Engineering Zhejiang University of Technology, Hangzhou 310014, P. R. China*
[‡]*fangfwu@zjut.edu.cn*
[§]*gcscaoxh@zjut.edu.cn*

Aqueous zinc-ion batteries (AZIBs) are the most promising candidates for large-scale energy storage devices due to the advantages of low cost, high safety, environmental friendliness and high energy density. However, the low Coulombic efficiency (CE) and short cycle life of AZIBs caused by dendrite growth, hydrogen evolution reaction and corrosion of Zn anode, are limited the development and application of AZIBs in the future. To solve these problems, many works focused on the modification of Zn anode and electrolyte optimization have been widely reported. The separator–electrolyte interface and the separator–anode interface play a significant part in the behavior of zinc ions. Owing to the importance of separators for batteries, this paper reviews the requirements and optimization strategies of separators for AZIBs. It is mainly based on the surface modification of conventional separators (e.g. glass fiber, cellulose separators), the introduction of an intermediate membrane in the interlayer of the separator and anode, and the preparation of new-type separators to replace the conventional

---

[¶]Corresponding authors.
To cite this article, please refer to its earlier version published in the Functional Materials Letters, Volume 16(8), 2340017 (2023), DOI: 10.1142/S1793604723400179.

separators. In addition, this review proposes a further outlook on the future development of separators for AZIBs.

Keywords: Aqueous zinc-ion batteries; separators; surface modification; intermediate membrane; new-type separators.

1. Introduction

Because of the over-exploitation of fossil fuels and increasing environmental pollution problems, the development of clean energy like solar, wind and tidal energy, naturally, is receiving increasing attention.[1-5] Energy storage systems (ESS) play an essential role in realizing the regulation of the power system, balancing user demands and improving energy efficiencies.[6,7] Currently, lithium-ion batteries (LIBs) still dominate the energy market, especially in electric vehicles and portable devices.[8,9] However, the large-scale utilization of LIBs has been hampered by the high price, restricted lithium resources and hidden danger of organic electrolytes.[10-12] It is an urgent task to develop the EES with high safety, low cost and high energy density. Several highly safe multivalent metal ion batteries have been recommended and extensively researched, such as Zn^{2+}, Ca^{2+}, Mg^{2+} and Al^{3+}-based batteries.[13-19] Among them, aqueous zinc-ion batteries (AZIBs) have been considered as a promising candidate for ESS because of the high safety, low cost (around \$65/kWh of AZIBs versus \$181/kWh of LIBs),[20] abundant zinc resource, high anodic theoretical capacity (820 mAh g^{-1}), low polarization and low redox potential (−0.76 V versus standard hydrogen electrode).[21-23]

Regrettably, the development of AZIBs is plagued by the dendrite growth and side reaction (corrosion and hydrogen evolution reaction) of Zn anode, the structural instability of cathode materials, resulting in a low Coulombic efficiency (CE) and poor cycling stability of AZIBs.[24-26] The uncontrolled Zn dendrites easily puncture the separator, leading to the short circuit of the battery, and the dislodged dendrites can form "dead Zn".[27,28] Moreover, the side reactions of Zn anode usually consume electrolytes and form a passivation layer on anodic surface.[29-31] Numerous approaches have been brought out to address these issues from the perspective of electrode/electrolyte design, including surface modification of Zn anode,[32-34] structural design of Zn anode,[35,36] electrolyte optimization[37-39] strategy. In fact, the separator acts as an insulating layer between the anode and cathode in batteries to prevent short circuits of batteries, as well as a carrier of electrolytes and ion transport channels. It is also important for the whole battery performance.[40,41] The micropores in the separator facilitate the passage of Zn^{2+}. Based on the previously reported findings, the ideal separators should have the following qualities: (1) Good electrical insulation: conducting ions but not electrons, and ensuring the mechanical isolation of anode and cathode, avoiding short circuit of the batteries; (2) Excellent chemical stability: chemically nonreactive with electrode and electrolyte during the cycle of the batteries; (3) High mechanical strength: making sure not easily puncture by dendrites; (4) Appropriate pore properties: including pore size and porosity, the uniform and suitable pores size distribution of the separator can regulate the uniform transmission of ions on electrodes, thereby extending batteries life; (5) Good electrolyte wettability: to ensure sufficient contact of the separator and electrolyte, increasing ionic conductivity and facilitating ion transport; (6) The appropriate thickness: the thickness of separator is related to the internal resistance and energy density of whole battery, so it should be reduced as thin as possible.

Currently, commercial glass fiber (GF) is frequently utilized as a separator of AZIBs owing to its high ionic conductivity and good aqueous electrolyte wettability.[42,43] Besides, GF also has good corrosion resistance and low coefficient of thermal expansion (resistant to 550°C), which is in favor of the maintenance of stable performance for batteries under high voltage. However, the low tensile strength and uneven pores of the GF separator cause the slow and uneven transport of Zn^{2+}, leading to the uneven deposition of Zn and dendrite growth on the Zn anode surface and further puncturing the separator.[44] Commercial cellulose membranes are also widely used as separators for AZIBs due to the advantages of strong hydrophilicity and low price. Nevertheless, the irregular pore structure and poor mechanical properties of commercial cellulose separators usually aggravate the uneven Zn deposition.[45]

Compared with traditional separators (commercial GF and cellulose separators), functional separators can be beneficial to enhance the performance of the whole battery by adjusting ion diffusion,[46-48] guiding the selective migration of ions,[49,50] or homogenizing electric field distribution.[42,51] Aware of the influence of functional separators, research on separators has increased in the past five years. Nevertheless, the separators of AZIBs have received less attention compared to anode and electrolyte (Fig. 1), and few reviews have been reported on the progress of separator research. Therefore, in this review, we focus on separators for AZIBs, including GF-based separators, cellulose-based separators and others. We systematically review the recent developments of the separators based on the following perspective of separators optimizing strategies (Fig. 2): (1) Modification of conventional separators (e.g. GF and cellulose separators). The functional modified layer can be considered an electrode extension to regulate Zn uniform deposition/stripping. (2) Introduction of an intermediate membrane.

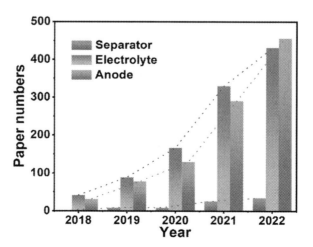

Fig. 1. Statistical chart of paper numbers related to separator, electrolyte and anode of AZIBs from 2018 to 2022. The data was from the Web of Science.

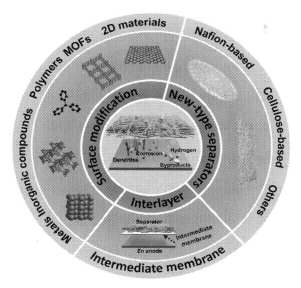

Fig. 2. An overview of optimization strategies for separators of AZIBs.

The construction of new interface was carried out by introducing intermediate membrane between anode and separator. The interface can regulate the kinetic behavior of Zn^{2+}, and solve the peeling and uniformity of the modified layer on the separator. (3) Other separators. The preparation of new types of separators with uniform pores and high mechanical strength replace the conventional separators of AZIBs, such as nafion-based separator and cellulose-based separators.

2. The Optimization Strategies of Separators for AZIBs

2.1. *Surface modification of traditional separators*

Traditional separators (e.g. GF and cellulose separators) are widely used in AZIB, but the inherent property of their large inhomogeneous internal porosity will lead to uneven Zn deposition and intense dendrite growth.[43] Moreover, the electric field at the interface is inhomogeneous, leading to inhomogeneous Zn^{2+} flux. Therefore, researchers proposed the functionalized separators were prepared by surface modification with different materials. The functionalized layer can promote uniform deposition of Zn^{2+} by homogenizing the interfacial electric field,[49] changing the surface structure of the separator,[52] and modulating the ionic behaviors.[50] Hence, the cycle life of the Zn anode is significantly extended. A lager number of materials have been explored to modify the separators for high-performance AZIBs.[42,51,53] Herein, we summarize the modification materials based on the structural features, including two-dimensional (2D) materials (graphene, MXene), metal–organic frameworks (MOFs), polymers, inorganic compounds and metallic materials.

2.1.1. *Two-dimensional (2D) materials*

2D materials usually have open ion transport channels, large specific surface area, and adjustable surface functional groups, which are widely used in the electrode design and separator modification.[56–58] As a typical 2D material, graphene has a large specific surface area, good chemical stability and high conductivity characteristics. The graphene layer on the surface of the separator can form the conductive support and homogenize the electric field distribution, which promotes rapid ion transport.[56,59] Li et al. developed a vertical graphene (VG)-modified separator by growing VG on the GF surface via plasma-enhanced chemical vapor deposition (PECVD). Subsequently, the surface of the VG-modified separator was doped by O and N atoms through the plasma method (Fig. 3(a)).[42] The separator consisted of VG coated on interweaving GFs (Fig. 3(b)), and the surface of GF was uniformly covered by VG arrays. Three-dimensional (3D) conductive VG scaffold could uniform the electric field on anodic surface (Fig. 3(c)), reduce the local current density, delay the initial formation point of Zn dendrite and inhibit dendrite growth. Furthermore, the O/N co-doped graphene enhanced the adsorption of Zn^{2+} and promoted the uniform deposition/stripping of Zn. Therefore, $Zn\|V_2O_5$ using VG-modified separator achieved a high capacity retention of 75% at 5 A g^{-1} after

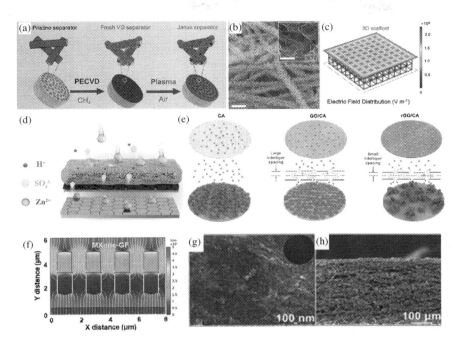

Fig. 3. (a) Schematic of synthesis process of the VG-modified separator. (b) Top-view SEM image of the VG-modified separator. (c) Electric field distribution of the VG-modified separator case (3D scaffold structure). Reproduced with permission from Ref. 42. Copyright 2020, Wiley-VCH. (d) Schematic diagram of regulating the transport behavior of Zn^{2+}, SO_4^{2-} and H^+ by the Janus separator with ion selectivity. Reproduced with permission from Ref. 49. Copyright 2022, Wiley-VCH. (e) Schematic of Zn^{2+} deposition using CA, GO/CA and rGO/CA separators, respectively. Reproduced with permission from Ref. 54. Copyright 2021, American Chemical Society. (f) The simulated electric field distribution of the Janus MXene-GF separator. Reproduced with permission from Ref. 53. Copyright 2022, Wiley-VCH. (g and h) SEM images of MXene@NiO-modified separator (illustration is digital photos of the separator). Reproduced with permission from Ref. 55. Copyright 2022, American Chemical Society.

1000 cycles, better than that of $Zn\|V_2O_5$ using GF. To achieve ion selectivity of the separator, Zhang *et al.* proposed the construction of Janus separators with ion selectivity by adsorbing sulfonated cellulose on graphene sheets, and modified on GF using spin coating technique.[49] The Janus separator was negatively charged due to the sulfonic acid group, which repelled SO_4^{2-} in the electrolyte. Moreover, the hydroxyl group of the sulfonic acid cellulose anchored H^+ to form hydrogen bonds. Benefiting from the above advantages, the Janus separator effectively inhibited corrosion and hydrogen evolution reaction of Zn anode (Fig. 3(d)). As a result, Zn∥Zn symmetric cell with the Janus separator cycled stably for more than 1400 h at a high current density of 10 mA cm^{-2} and a high area capacity of 10 mAh cm^{-2}.

Compared to graphene, abundant zincophilic O-containing groups of graphene oxide (GO) can supply a significant number of nucleation sites with low interfacial energy, enabling uniform Zn deposition. Luo et al. reported the composite separator was constituted via GO nanosheet-modified cellulose acetate (CA) by a simple filtering method,[54] and the mass loading was only 4 μg cm^{-2}. H_2O could insert into the GO layer due to the hydrophilic O-containing functional groups of the GO, the larger layer spacing of GO allowed the migration of ions. The O-containing functional groups of GO provided abundant nucleation sites for uniform nucleation and growth of Zn. In addition, CA/GO could guide the deposition of Zn^{2+} along the (002) crystal plane because of a low lattice mismatch (about 7.4%) of GO with Zn (002) (Fig. 3(e)). Nonetheless, the oxidation of GO to reduced GO (rGO) removed most of the O-containing functional groups from the surface, so rGO exhibited reduced layer spacing and nucleation sites (compared to GO), leading to dendrite growth. Therefore, Zn|Zn cell with a CA/GO separator cycled stably for 500 h at 10 mA cm^{-2} and 1 mAh cm^{-2}.

Different from the single component of graphene, MXene as the innovative 2D material is composed of numerous transition metals, carbon, nitrogen and carbon–nitrogen compounds, and it is first reported by Naguib et al. in 2011.[60,61] MXene can be expressed as $M_{n+1}X_nT_x$ (n = 1–3), where T_x refers to the generated end groups during the etching process, such as –OH, –O, –F, etc.[62,63] In addition, MXene has become a hot research topic for the cathode materials of AZIBs due to its good conductivity. The conductivity of film with a thickness of only 214 nm is about 15,100 S cm^{-1}.[64,65] providing an idea for MXene modification of the separator and anode interface. Su et al. developed a Janus MXene-GF separator by spraying Ti$_3$C$_2$T$_x$ MXene nanosheets on GF surface.[53] The high dielectric constant of MXene-GF could construct a directed built-in electrical field with 50% increase in intensity via Maxwell–Wagner effect to hasten Zn^{2+} transfer dynamics (compared to GF) (Fig. 3(f)). Moreover, the abundant surface functional groups (–OH and –F) of MXene-GF separator attracted Zn^{2+} and repelled anions in the electrolyte, which further facilitated the transport process of Zn^{2+}. Zn||Zn symmetric cells with the MXene-modified GF separators achieved stable cycling for 1180 h at 1 mA cm^{-2} with 1 mA h cm^{-2}. To optimize the pores structure of the separator surface, a MXene/nanoporous NiO heterostructure engineered separator was designed.[55] The nanoporous NiO effectively reduced the local current density owing to its large surface area and high porosity. Besides, nanoporous NiO covered the uneven distribution of large pores of cellulose separator (Figs. 3(g) and 3(h)). Meanwhile, the zincophilic MXene@NiO uniformed the distribution of the electric field, reduced the nucleation overpotential of Zn deposition, achieved uniform deposition of Zn^{2+} and suppressed the occurrence of the side reactions of anode. Benefitting from the effect of the MXene@NiO-modified separator, Zn anode achieved a stable cycling of 500 h during Zn deposition/stripping at 10 mA cm^{-2} with 10 mAh cm^{-2}.

2.1.2. Metal-organic framework materials

MOFs are crystalline materials consisting of metal nodes connected with organic ligands, first reported by Yaghi in 1995.[66] MOFs are widely used in AZIBs because of their large specific surface area and high porosity, which facilitate electrolyte permeation and ions transport. Moreover, their structure and morphology can be controlled.[67–69] Among them, the UiO-66 and MOF-808 are zirconium (IV)-based MOFs materials (Zr-MOFs), which exhibit high porosity, and excellent chemical and thermal stability. In addition, the particle size of Zr-MOFs is adjustable, and appropriate particle size can be displayed by changing the junction and unit to achieve rapid ion migration and uniform Zn deposition.[70,71] Song et al. obtained a UiO-66-GF separator through UiO-66 in situ modification on GF under the hydrothermal method (Fig. 4(a)).[72] Because of the rich pore structure and large specific surface area of UiO-66 (990.3 m^2 g^{-1}), the electrolyte penetrated uniformly into UiO-66-GF, and the local current density near the Zn anode were reduced. Furthermore, X-ray diffraction (XRD) spectrum of Zn anodes of Zn∥MnO$_2$ full cells after cycling revealed that the anodic surface of Zn with UiO-66-GF separator exhibited a higher peak intensity ratio of Zn (002)/Zn (101) than that with GF. UiO-66-GF separator guided Zn^{2+} to preferentially deposit on (002) crystal planes of Zn, resulting in Zn deposition with a flat surface. In addition, the lower adsorption energy of H atoms and Zn (002) crystal planes than that of Zn (100) and Zn (101) crystal planes were obtained (Fig. 4(b)). The weakened H adsorption on the Zn (002) crystal plane contributed to improving corrosion resistance and inhibiting hydrogen evolution reaction on anodic surface. Thanks to the UiO-66-modified GF separator, the Zn∥Zn symmetric cell could cycle stably for 1650 h at 2 mA cm^{-2} with 1 mAh cm^{-2}. Wang et al. modified hydrophilic nonwoven separators using MOF-808 and rGO.[73] The excellent anionic sub-nanotunnels of MOF-808 could achieve uniform Zn^{2+} flux. The conductive rGO layer provided additional electron pathways and reduced the redox energy barrier of Zn at the interface. Hence, the "dead Zn" in the cycle was digested, and the Zn∥Cu cell achieved a CE of about 99.2% after 100 cycles at 2.0 mA cm^{-2} with 1 mAh cm^{-2}.

2.1.3. Polymers materials

Polymers usually exhibit excellent chemical stability, lightweight, and polymers have a wide variety.[74] Their flexible backbone can change the surface structure of the separator and effectively prevent

A review on advanced optimization strategies of separators 29

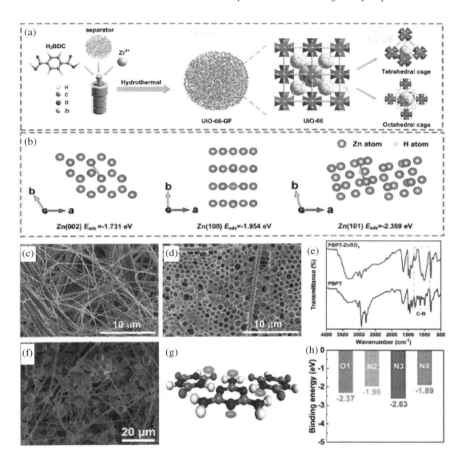

Fig. 4. (a) Schematic diagram of the synthesis of UiO-66-GF. (b) The adsorption energy of different crystalline planes of Zn with H atoms. Reproduced with permission from Ref. 72. Copyright 2022, Springer Nature. Top-view SEM images of (c) GF separator and (d) GF/PBPT-42% separator. (e) FTIR spectra of PBPT membranes immersed and not immersed in $ZnSO_4$ electrolyte. Reproduced with permission from Ref. 52. Copyright 2022, Elsevier. (f) SEM image of SM@GF separator. (g) Simulation of possible adsorption sites of Zn atoms with supramolecular structures and (h) their corresponding adsorption energy calculation results. Reproduced with permission from Ref. 48. Copyright 2021, Elsevier.

the puncture of dendrites.[52] More importantly, the polymer is rich in zincophilic functional groups, and the groups can provide abundant sites for Zn^{2+}, thus trapping Zn^{2+} and guiding the uniform distribution of Zn^{2+}.

Ether-bond-free aryl polymers have good mechanical properties and chemical stability, and abundant functional groups of their main and side chains provide continuous conductive channels for ions.[52,75–77] In view of these, poly(biphenyl piperidine triphenylmethane) (PBPT) was synthesized and then modified on the surface of GF to obtain GF/PBPT separator.[52] The PBPT possessed abundant secondary pore structure on GF surface, as shown in Figs. 4(c) and 4(d). The pore size and structure were adjusted by adding different amounts of PBPT polymers. The suitable pores of PBPT provided favorable transport channels for Zn^{2+} and imparted excellent mechanical stability to the GF/PBPT separator. Additionally, the interaction between PBPT and Zn^{2+} was detected using Fourier transform infrared (FTIR) spectra of dried PBPT membranes (with and without $ZnSO_4$) ($PBPT–ZnSO_4$ and PBPT) (Fig. 4(e)). Compared to PBPT, $PBPT–ZnSO_4$ significantly changed the C–N stretching vibration peak position and intensity in the range of 1324–841 cm^{-1}, which was related to the affinity between Zn^{2+} and tertiary amines (R_3N_s) in PBPT. The high affinity of PBPT and Zn^{2+} facilitated the adsorption and transference of Zn^{2+} to achieve uniform Zn deposition. Therefore, Zn∥Zn symmetric cells with PBPT-modified GF separator cycled for 1540 h at 0.5 mA cm^{-2} with 0.5 mAh cm^{-2}. Functional supramolecules are another type of polymer with abundant functional groups. Liu et al. reported SM@GF separator by modifying supramolecule on GF separator via filtration.[48] Scanning electron microscopy (SEM) image of SM@GF separator showed supramolecular growth on GFs (Fig. 4(f)). The binding energies of four different sites in the supramolecule structure with Zn atom were −2.37, −1.96, −2.36 and −1.89 eV (Figs. 4(g) and 4(h)), respectively. Thus, the supramolecule displayed an excellent affinity for Zn^{2+}, Zn^{2+} preferred to these zincophilic sites due to strong interactions, so a uniform Zn^{2+} flux could be achieved.

Zn^{2+} was uniformly deposited on the anode surface during deposition, inhibiting the formation of Zn dendrites. As a result, a stable Zn deposition/stripping for 2000 h in the Zn‖Zn symmetric cell at 1 mA cm^{-2} with 1 mA h cm^{-2} was obtained.

2.1.4. Inorganic compounds

Inorganic compounds (such as oxides and inorganic salts) are also widely used as materials for separator modification, regulating the uniform deposition of Zn^{2+} by homogenizing electric field distribution and regulating of Zn^{2+} flux. Yang et al. synthesized OH-terminated SiO_2 nanospheres, and then OH-terminated SiO_2 nanospheres were modified on the cellulose separator using drop-casting (Fig. 5(a)).[50] The SiO_2 nanospheres uniformly covered the cellulose surface (Fig. 5(b)). The interspaces between SiO_2 nanospheres provided orderly transport channels for Zn^{2+}, and homogenized the current density of separator surface (Fig. 5(c)). Meanwhile, the negative charge of OH-terminated SiO_2 adsorbed Zn^{2+} to homogenize Zn^{2+} flux with the pH change. Zn‖Cu asymmetric cell with the OH-terminated SiO_2-modified cellulose separator maintained a high CE of 99.62% after 2000 cycles at 1 mA cm^{-2} with 0.25 mA h cm^{-2}. Similarly, SiO_2 with hydroxyl groups and micro-mesoporous ion channels (IS/SiO_2–OH) was modified on GF separator.[78] The IS/SiO_2–OH layer homogenized the electric field distribution on the GF surface, and the micro-mesoporous channels were able to adjust the Zn^{2+} flux. In addition, the low zincophilic of IS/SiO_2–OH facilitated the preferential deposition of Zn away from the surface of the separator, and prevented from puncturing the separator. Benefiting from IS/SiO_2–OH layer, the capacity retention of Zn‖NVO full cell was 70.7% at 2 A g^{-1} after 1500 cycles, while the Zn‖NVO cell with GF separator failed at about 300 cycles. Modifying the separator should consider the synergistic effect of the interface between the separator and electrolyte or separator and anode. Liang et al. proposed dual-interface engineering by modifying $BaTiO_3$ (BTO) nanoparticles on the GF surface and filling the

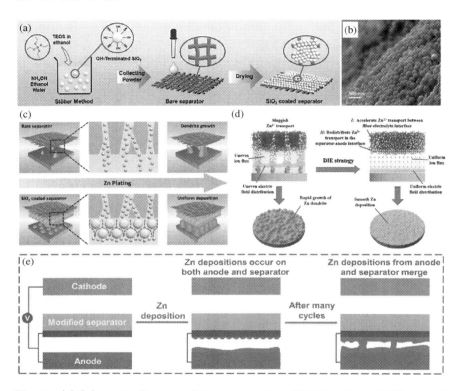

Fig. 5. (a) Schematic diagram of the preparation of OH-terminated SiO$_2$-coated separator. (b) Cross-section SEM image of OH-terminated SiO$_2$-coated separator. (c) Schematic diagram of Zn deposition based on cellulose separator and the SiO$_2$-modified cellulose separator. Reproduced with permission from Ref. 50. Copyright 2022, American Chemical Society. (d) Mechanism scheme of dendrite-free Zn deposition achieved by the dual-interface engineering of the separator. Reproduced with permission from Ref. 44. Copyright 2022, Wiley-VCH. (e) Schematic diagram of Zn deposition in the contact area between the Sn-modified separator and the anode. Reproduced with permission from Ref. 51. Copyright 2021, Springer Nature.

interior of GF.[44] The separator played an accelerator of Zn^{2+} and homogenized the distribution of Zn^{2+} on the separator–anode interface because of the spontaneous polarization effect and zincophilicity of BTO. Therefore, the reversibility of Zn^{2+} deposition/stripping were enhanced (Fig. 5(d)). Benefitting from the BTO modification, Zn‖Zn batteries maintained 1600 h at 10 mA cm^{-2} with 2.5 mAh cm^{-2}.

2.1.5. Metallic materials

In addition to the modification materials above, the metal coating is also considered as an efficient way to modify the separator. Metals usually possess high electrical conductivity, which is common in the protection of anode.[79,81] The introduction of conductive modification layer between anode and separator is hopeful to regulate the electric field on the surface of Zn anode. Hou *et al.* reported the cellulose separator was modified by Sn metal layer with high conductivity and zincophilicity through magnetron sputtering method.[51] The Sn-coated separator displayed a uniform electric field due to the Sn conductive layer on the separator. Besides, the uniform electric field promoted uniform deposition of Zn^{2+} on the anode. The Zn^{2+} flux under this uniform electric field was controlled owing to the zincophilic of Sn layer. During the Zn deposition process, Zn^{2+} deposited toward the Sn layer, inhibiting the unidirectional vertical deposition of zinc from piercing the separator (Fig. 5(e)). The Zn‖Zn cell with Sn-coated cellulose separator displayed an extremely long cycle time of 3800 h at 2 mA cm^{-2} with 2 mAh cm^{-2}. Unfortunately, the report about metal modification layer on separator is far more than enough, maybe restricted by the difficulty of synthesis technology. Thus, it is necessary to develop the metal layer modified on the separator via a simple and appropriate method.

2.2. Introduction of an intermediate membrane between separator and anode

Despite the exhilarating effects of the separator modification, the inhomogeneity and peeling issues of the modified layer on the separator surface will appear during the cycling process. One efficient method to solve the above issues is that an intermediate membrane is introduced between the separator and the Zn anode. The intermediate membrane can form a physical shielding layer at the separator–anode interface to avoid piercing the separator by dendrite. The intermediate membrane always exhibits excellent stability and uniform pore structure, which creates the advantage for the uniform distribution of Zn^{2+}.

Liang et al. prepared N, O co-doped carbon nanofiber (CNF) interlayer between anode and separator by the electrostatic spinning technique as shown in Fig. 6(a),[82] which induced uniform nucleation and deposition of Zn^{2+} by adding zincophilic sites at the

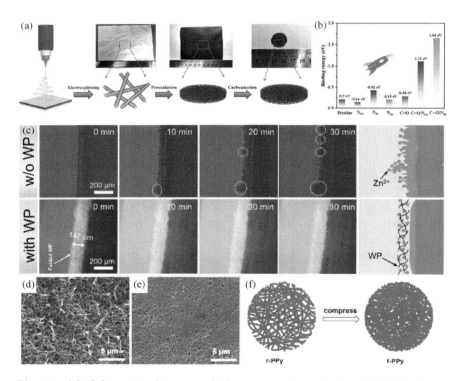

Fig. 6. (a) Schematic diagram of the preparation of the CNF interlayer. (b) The binding energies of Zn atoms with different adsorption sites on CNF. Reproduced with permission from Ref. 82. Copyright 2021, Elsevier. (c) *In situ* optical microscope images of Zn anode with or without WP interlayer at a current density of 5 mA cm^{-2} and the corresponding schematic diagram. Reproduced with permission from Ref. 83. Copyright 2022, Elsevier. SEM images of (d) r-PPy and (e) f-PPy. (f) Illustration of compression process from r-PPy to f-PPy paper. Reproduced with permission from Ref. 85. Copyright 2022, Elsevier.

interface. The binding energies of Zn atoms with N and O doping sites were calculated by density functional theory (DFT), and the binding energies of double doping sites C=O/N$_{Pd}$ and C=O/N$_{Pr}$ with Zn became significantly higher compared with the undoped sites (Fig. 6(b)). The high affinity to Zn of CNF interlayer provided abundant active sites for Zn and reduced the nucleation potential of Zn, easily trapping more Zn^{2+}. The high Zn^{2+} transfer number (t_{Zn}^{2+}) indicated that the CNF interlayer effectively reduced the concentration difference near interface and alleviated the polarization problem. Due to the introduction of a CNF interlayer, the Zn∥Zn cell cycled stably for 1200 h at 5 mA cm^{-2} and 1 mAh cm^{-2}, and the overpotential was only 59.5 mV.

Weighing paper (WP) is the most common item in the laboratory. Guo et al. demonstrated that WP is an outstanding interlayer to stabilize Zn anode.[83] The WP displayed a smaller pore than that of filter paper, owing to the insertion of glycerol into the pores of WP in the production process. Besides, zincophilic-O groups of the WP contributed in providing much more active sites for the adsorption of Zn^{2+}. The tensile strength of WP was 42 MPa, much higher than GF (0.3 MPa). Benefiting from the 3D structure and zincophilic-O groups of WP, the functional WP interlayer adjusted the mobility of Zn^{2+} and prevented immediate contact from anode and electrolyte. Furthermore, the desolvation process of [Zn(H$_2$O)$_6$]$^{2+}$ was promoted and the distribution of Zn^{2+} was regulated by this WP interlayer. As shown in Fig. 6(c), in situ optical microscopy images of Zn anode with WP interlayer showed uniform Zn deposition at different times. The cycle life of Zn∥Zn cell with WP interlayer remained stable for 2400 h at 1 mA cm^{-2} with 1 mAh cm^{-2}.

Polypyrrole (PPy) is a typical conductive polymer with abundant hydrophilic –NH functional groups and high wettability, and thus is supposed to use as an advanced zincophilic interlayer.[84,85] PPy paper was prepared by PPy-coated nanocellulose via a simple chemical polymerization method.[85] The PPy paper exhibited interconnected nanofibrous structure and narrow mesoporous feature (Figs. 6(d)–6(f)), as well as high wettability and good affinity to Zn^{2+}. The mesoporous PPy paper was utilized as interlayer between anode and separator, which could improve the stability of Zn anode. Compared with the pristine PPy paper (r-PPy), the compressed PPy (f-PPy) displayed flat surface and thinner thickness (17 μm), easily kept in close contact with Zn anode and promoted the uniform transport of Zn^{2+}. Besides, the interfacial electric field distribution of anode was homogenized by this f-PPy functional interlayer, owing to its good conductivity. The Zn∥Zn cell with f-PPy interlayer maintained a stable cycle for 1600 h at 2 mA cm^{-2} with 2 mA h cm^{-2}.

2.3. Preparation of new-type of separators

Although the performance of AZIBs is notably improved by surface modification of the separator and the introduction of the intermediate membrane. Nevertheless, these strategies do not play a crucial impact in changing the intrinsic structure of the separator, and increase the cost of the separator. Beyond that, optimizing the structure and components of the separator by preparing new-type separators is a hot research topic. The new-type separators (such as Nafion and cellulose separators) usually have excellent mechanical strength, uniform pore structure.

2.3.1. Nafion-based separators

Nafion consists of a hydrophobic main chain and a hydrophilic side chain with rich sulfonic acid group. The negatively charged sulfonic acid group of side chain gives Nafion cation selectivity.[86] Moreover, the high mechanical strength can improve the dendrite puncture resistance of Nafion membrane.[87,88] Hence, the Nafion-based separators are considered as a suitable separator for AZIBs.

The most common method is to immerse the Nafion membrane in a salt solution in order to replace the proton of side chains with cations. Wu et al. reported that Nafion membrane was soaked in $ZnSO_4$ electrolyte for three days so that Zn^{2+} replaced the protons of Nafion side chain end groups, and finally the recoverable Zn-Nafion separator was obtained.[86] The surface of Zn-Nafion separator was

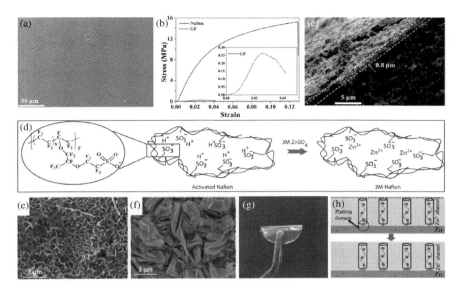

Fig. 7. (a) SEM image of Zn-Nafion separator. (b) Stress–strain curves of Zn-Nafion separator and GF separator (c) SEM image of the SEI layer generated on the cathode surface in a Zn|Zn-Nafion|V$_2$O$_5$ full cell. Reproduced with permission from Ref. 86. Copyright 2021, Royal Society of Chemistry. (d) Schematic representation of the Zn^{2+} exchanged Nafion. (e) Morphology of Zn anode after 1300 cycles of Zn∥V$_2$O$_5$ with Zn^{2+}-integrated Nafion separator. (f) Morphology of Zn anode after 650 cycles of Zn∥V$_2$O$_5$ with GF separator. Reproduced with permission from Ref. 88. Copyright 2019, Wiley-VCH. (g) Digital photo of ZPSAM. (h) Schematic diagram of Zn deposition on Zn anode using ZPSAM. Reproduced with permission from Ref. 89. Copyright 2020, Elsevier.

dense and smooth (Fig. 7(a)). In Fig. 7(b), the stress–strain curves showed that Young's modulus of the Zn-Nafion separator (388.53 MPa) was about 20 times higher than that of the GF separator (18.14 MPa). The excellent mechanical strength of Zn-Nafion separator could effectively inhibit the dendrite from piercing the separator. Additionally, the Zn-Nafion separator could homogenize the electric and Zn^{2+} concentration fields to obtain uniform Zn^{2+} deposition at the separator–anode interface. The Zn∥V$_2$O$_5$ full cell with the Zn-Nafion separator formed a solid–electrolyte interphase (SEI) layer with a thickness of 0.8 μm on the cathode surface during cycling process (Fig. 7(c)), inhibiting the dissolution of cathode and improving the electrochemical performance of AZIBs. Hence, Zn|Zn-Nafion|V$_2$O$_5$ offered a high capacity of 414.1 mAh g^{-1} at 1 A g^{-1} after 250 cycles with a high capacity retention of 96.8%. More importantly, the Zn-Nafion separator could be recycled at least 10 times, but GF could only be reused three times. The recyclability of Zn-Nafion separator reduces its cost to 47.3%, which is cheaper than GF separator. A similar effect in the Zn^{2+}-integrated Nafion separator was prepared by soaking Nafion 212 in ZnSO$_4$ (Fig. 7(d)).[88] The interaction between the electron-rich sulfonic acid group (–SO$_3^-$) and Zn^{2+} in Nafion could promote the dissociation of ZnSO$_4$, and facilitate the uniform Zn^{2+} deposition, inhibiting the diffusion of anions to the anode. Moreover, the Zn^{2+}-integrated Nafion separator could provide high $t_{Zn^{2+}}$ and low activation energy of ionic conduction (0.069 eV) to reduce the deposition/stripping potential of Zn. Furthermore, the Zn anodic surface of Zn||V$_2$O$_5$ with GF generated sharp deposits after cycling, but unique thin nanowall-like deposits were produced on the anodic surface using a Zn^{2+}-integrated Nafion separator (Figs. 7(e) and 7(f)), which could promote the long cycling performance of AZIBs. Zn||V$_2$O$_5$ with Zn^{2+}-integrated Nafion separator showed capacity retention of 84% at 5 A g^{-1} after 1300 cycles, while the capacity retention of Zn||V$_2$O$_5$ cell using GF separator only was 78% after 650 cycles. Similarly, a zincic perfluorinated sulfonic acid membrane (ZPSAM) separator with semi-immobilized anions was designed by immersing activated perfluorosulfonic acid membranes in zinc acetate that Zn^{2+} replaced protons (Fig. 7(g)).[89] Zn^{2+} could be transported in negatively charged channels in the ZPSAM (Fig. 7(h)), and SO$_4^{2-}$ transport could be suppressed, thus inhibiting the generation of by-products on the surface of anode. Moreover, the ZPSAM separator allowed Zn^{2+} to achieve nano-wetting contact at the electrolyte–anode interface, avoiding "the tip effect" and inhibiting the growth of dendrites. Therefore, the Zn||Zn with a ZPSAM separator with a lifespan of 2000 cycles at 2 mA cm^{-2} with 0.5 mAh cm^{-2}.

2.3.2. Cellulose-based separators

Cellulose has become one of the ideal choices for the separator of AZIBs thanks to its good insulating properties, excellent hydrophilic properties, environmental friendliness, and abundant natural resources on earth. It can be extracted from cotton, bamboo and microorganisms (such as fungi).[90-92] However, the irregular pore structure of traditional commercial cellulose separator may aggravate uneven Zn deposition.[45]

As shown in Fig. 8(a), Zhou *et al.* removed substances other than cellulose from cotton by NaOH and H_2O_2, and prepared a cellulose film (CF) separator via filtration.[93] Unlike the inhomogeneous macropores of GF, the CF separator had dense and uniform

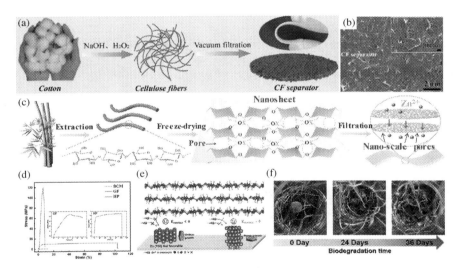

Fig. 8. (a) Schematic diagram of the preparation of CF separator. (b) SEM image of CF separator. Reproduced with permission from Ref. 93. Copyright 2021, Elsevier. (c) Schematic diagram of the preparation of bamboo cellulose separator. Reproduced with permission from Ref. 98. Copyright 2022, Elsevier. (d) Tensile curves of the BCM and GF separator. (e) Schematic representation of the crystallographic orientation of BCM separator manipulated Zn deposition. (f) Biodegradation test of the BCM separator. Reproduced with permission from Ref. 99. Copyright 2022, Elsevier.

pores (Fig. 8(b)). The uniform and micropores could homogenize the Zn^{2+} flux and allow Zn^{2+} to uniformly deposit on the Zn anode surface. The excellent tensile strength of the CF separator was 29.2 MPa, 34 times that of the GF separator. The hydroxyl groups in the CF separator could form strong hydrogen bonds with H_2O, thus reducing the desolvation activation of $Zn(H_2O)_6^{2+}$ from 43.4 kJ mol^{-1} to 33.6 kJ mol^{-1}, and the nucleation overpotential of Zn was reduced to 48.4 mV (compared to 70.1 mV in GF). Impressively, Zn||Zn symmetric cell with the CF separator maintained stable cycling for 2000 h at 1 mA cm^{-2} with 1 mA h cm^{-2}. Apart from cotton, cellulose extracted from bamboo exhibits high mechanical strength, excellent electrical insulating properties, excellent hydrophilicity and good film-forming ability.[94,95] Moreover, bamboo is the fastest-growing plant with abundant natural resources.[96,97] Fu et al. reported a bamboo cellulose separator extracted from bamboo via self-assembly method (Fig. 8(c)).[98] The tensile strength of the bamboo cellulose separator was 81 MPa, which was much higher than that of GF (0.54 MPa). The high strength was conducive to inhibiting Zn dendrites. The natural nanoscale pores of bamboo cellulose separator could provide a convenient transport channel for Zn^{2+} and regulate an ordered and oriented Zn^{2+} flux, thus achieving uniform deposition of Zn. The Zn||Zn symmetric cell used bamboo cellulose separator and achieved an outstanding cycle life of 5000 h at 0.5 mA cm^{-2} with 0.25 mA h cm^{-2}. In addition to this, bacterial cellulose exhibits good hydrophilicity, excellent pore structure, mechanical properties and biodegradability. A biomass nanocellulose membrane (BCM) separator was constructed by a simple papermaking process.[99] The tensile test results showed that the tensile strength of BCM with a thickness of 9 μm was 120 MPa, while the GF could only withstand 0.45 MPa (Fig. 8(d)).[99] The DFT calculations revealed that the BCM displayed a high affinity to Zn (100), contributing to preventing the deposition of Zn^{2+} along the (100) crystal plane. However, the chemical interaction between the BCM and (002) was negligible, so Zn^{2+} could deposit on the anode toward (002) crystal plane (Fig. 8(e)). Zn|BCM|Zn symmetrical cell achieved a long cycle of 4000 h at 0.5 mA cm^{-2} with 0.1 mA h cm^{-2}. Experiments demonstrated that in the natural environment, this biomass cellulose could be degraded after 36 days, highlighting the environmental friendliness (Fig. 8(f)).

To enrich the functions of cellulose separators, some researchers introduced functional groups and materials for cellulose separators. Ge et al. replaced Na$^+$ in cellulose nanofiber sodium sulfonate (CNF-SO$_3$Na) with Zn^{2+} by ion exchange to obtain a single ionic nanofiber sodium sulfonate (CNF-SO$_3$Zn) separator.[45] SEM image and digital photo (shown in Figs. 9(a) and 9(b)) demonstrated that the surface of the CNF-SO$_3$Zn separator with micropores was smooth and dense. In addition, the thickness of the CNF-SO$_3$Zn separator was controlled within 25 μm, which was much thinner than the common GF separator (180 μm). Moreover, the CNF-SO$_3$Zn separator showed significantly higher t_{Zn}^{2+} (0.7) than other separators (e.g. 0.24 for filter paper, 0.15 for GF). What is more, the water activity was reduced due to hydroxyl and sulfonic acid groups of the CNF-SO$_3$Zn separator forming hydrogen bonds with H$_2$O, inhibiting hydrogen evolution reaction. Surprisingly, the batteries with the CNF-SO$_3$Zn separator can achieve a high depth of discharge and high-capacity utilization with a small amount of electrolyte. Cao et al. mixed ZrO$_2$ particles and polymethyl cellulose in deionized water, and obtained a cellulose ZrO$_2$-nanofibers composite (ZC) separator by solution casting.[100] The ZC separator with a thickness of 50 μm showed tiny pores of 10–50 nm, and the surface of the ZC separator was smooth and dense (as shown in Fig. 9(c)). In Fig. 9(d), the ZC separator could regulate electric field distribution to promote uniform Zn deposition and reject anions (such as SO$_4^{2-}$) attributed to the high dielectric constant of the ZrO$_2$ particles ($\varepsilon = 25$) dispersed in the separators, thus homogenizing Zn deposition and inhibiting the side reaction of anode. Zn||Zn symmetric cell using the ZC separator could be stably cycled for 2000 h at 0.5 mA cm^{-2} with 0.25 mA h cm^{-2}.

42 F. Du et al.

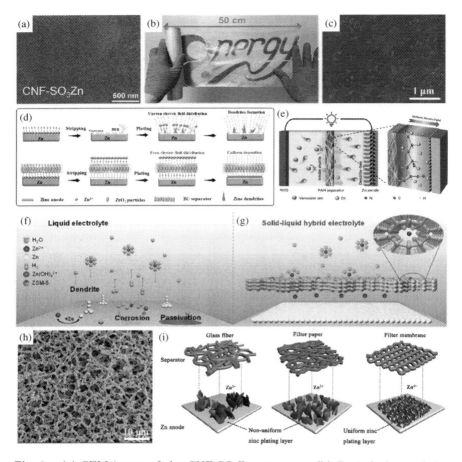

Fig. 9. (a) SEM image of the CNF-SO$_3$Zn separator. (b) Digital photo of the CNF-SO$_3$Zn separator. Reproduced with permission from Ref. 45. Copyright 2022, Wiley-VCH. (c) SEM image of the ZC separator. (d) Schematic diagram of Zn^{2+} deposition on Zn anode using cellulose and ZC separator. Reproduced with permission from Ref. 100. Copyright 2021, Elsevier. (e) The working mechanism of the PAN separator. Reproduced with permission from Ref. 101. Copyright 2022, Wiley-VCH. Schematic diagram of a typical reaction process on a Zn anode (f) without and (g) with a ZSM-5 separator. Reproduced with permission from Ref. 102. Copyright 2022, Wiley-VCH. (h) SEM image of the filter membrane. (i) Schematic diagram of the regulation of Zn deposition and dendrite growth using GF, filter paper and filter separator, respectively. Reproduced with permission from Ref. 43. Copyright 2020, Wiley-VCH.

2.3.3. Other separators

To date, a variety of separators have been reported. In addition to the above separators, some other separators of AZIBs have also the advantage of pore structure and chemical composition, achieving good results in modulating the behavior of ions. The membranes prepared by electrostatic spinning technology have the advantages of controlled composition and structure. Polyacrylonitrile (PAN) nanofiber separators with a thickness of 69 μm (675 μm for GF) were prepared by electrostatic spinning technique.[101] The PAN separator not only contained amounts of pores and a large surface area, but also high mechanical properties. The abundant zincophilic CN– groups of the PAN separator could regulate the Zn^{2+} flux (Fig. 9(e)). In addition, the PAN separator could homogenize the electric field and concentration field distribution of Zn^{2+} at the separator–electrode interface, promoting the uniform deposition of Zn^{2+} and inhibiting dendrite growth. The vanadium content of the anode side of the Zn|PAN|$NH_4V_4O_{10}$ cell gradually decreased, demonstrating that the shuttling of dissolved ions of cathode was effectively inhibited. Zn||$NH_4V_4O_{10}$ cell with a PAN separator achieved 84.3% capacity retention after 1000 cycles at 5 A g^{-1}, significantly higher than that of Zn||$NH_4V_4O_{10}$ cell using GF separator (57.8%). Furthermore, developing a separator with molecular sieve properties is an effective way to improve Zn deposition behavior. Zhu et al. reported the molecular sieve films (ZSM-5) of aluminosilicates.[102] 0.3–2.5 nm pore channels of ZSM-5 was thought to homogenize Zn^{2+} flux and guided uniform Zn^{2+} deposition. More importantly, as shown in Figs. 9(f) and 9(g), the ZSM-5 molecular sieve easily absorbed the free water molecules in the $ZnSO_4$ solution and the solvated water molecules around Zn^{2+} by the formation of hydrogen bonds between abundant O atoms in the ZSM-5 and H_2O. Therefore, the ZSM-5 significantly inhibited free water activity, and restrained Zn dendrites and side reactions of Zn anode.[102] With the protection of ZSM-5, the Zn||Zn cell could be stably cycled for 2000 h at 1 mA cm^{-2} with 1 mAh cm^{-2}.

44   F. Du et al.

Commercial cotton towels (CT) have excellent mechanical properties, high chemical stability, low price, high ionic conductivity advantages. The use of CT as a separator has been shown the enhanced electrochemical performance of zinc anode.[103] The excellent mechanical properties of the CT separator enhanced the puncture resistance of the separator. The strong interaction between the abundant hydroxyl groups of CT and Zn^{2+} could refine Zn grains and uniform Zn deposition. Therefore, the Zn∥Zn symmetric cell equipped with the CT separator cycled over 1200 h at 1 mA cm^{-2} and 0.5 mAh cm^{-2}. In addition, the low price of CT makes it possible to use it as a commercial separator. Qin et al. made the filter membrane with uniform pore distribution as a separator, studying the effect of pore structure of separator on Zn deposition.[43] The Zn deposition distribution of the three separators using GF, filter paper, and untreated filter paper during cycling was summarized. Filter separators could promote the uniform deposition of Zn^{2+} due to uniform micropores distribution (Figs. 9(h) and 9(i)).[43] As a result, Zn∥Zn symmetric cells equipped with the filter separators could achieve a long cycle of 2600 h with a low voltage hysteresis of 47 mV at 1 mA cm^{-2} with 1 mAh cm^{-2}.

3. Conclusion and Outlook

In conclusion, AZIBs have become the most promising candidates in ESS due to high safety, low cost, environmental friendliness and high energy density. However, Zn anode has encountered uneven Zn^{2+} deposition, Zn dendrite growth, corrosion and side reactions of Zn anode, seriously restricting the further development of AZIBs. Numerous works and reviews focused on surface modification of Zn anode and electrolyte design have been presented to solve these problems. There are an increasing number of papers on optimization strategies of separators, but few review papers on separators for AZIBs have been reported. Therefore, this paper reviews recent advances in separator optimization strategies, focusing on the advanced separators: (1) The development of modification of traditional separators. (2) The introduction of an intermediate membrane between the separator and Zn anode. (3) The new-type separator to replace the conventional separators of AZIBs (some parameters and properties of these separators are summarized in Table 1). In view of this, designing and preparing the novel separators are both urgent and challenging, and the following aspects should be considered:

(1) The cost and preparation process of functional separators need to consider the practical applications of AZIBs in the future. Although the modification of traditional separators and the introduction of an intermediate membrane can improve the performance of AZIBs, they will increase the cost of the batteries. Therefore, it is necessary to explore the design and preparation of new separators and cellulose-based separators with low cost and simple technology.

(2) Advanced *in situ* characterization techniques (*in situ* SEM, *in situ* XRD, etc.), theoretical calculation and simulation should be combined to study the influence of structural and interfacial properties of the separator on the Zn deposition/stripping behaviors from a microscopic perspective. The optimal range of separator structural parameters such as porosity, thickness, etc. should be obtained, which are conducive to achieving consistency in production design.

(3) Although novel separators can effectively guide the uniform deposition of Zn^{2+}, in most research works, the anode can only achieve long cycle life at low current densities and capacities. To meet modern society's demand for high-capacity and rapid charge/discharge functions, separators should be developed that can be charged and discharged for long periods at high current densities and capacities.

(4) Most studies only focus on the effect of the separators on the Zn anode, while the research on the effect on the cathode should be enhanced. The mechanism of the separator to promote ion diffusion and improve the stability of the cathode structure needs to be further explored.

(5) Fundamental understanding of the influence of electrolyte on separator for AZIBs is insufficient. In-depth research on electrolyte and separator is necessary. Therefore, the research of

Table 1. Summary of separator parameters and corresponding electrochemical performance of Zn||Zn, full cells.

Separator	Thickness/tensile strength	Ionic conductivity	Lifespan of Zn||Zn cells	Performance of full cells	Ref.
VG/GF	200 μm/—	—	600 h/10 mA cm^{-2} 1 mA h cm^{-2}	Zn||V$_2$O$_5$ (75% capacity retention at 5 A g^{-1} after 1000 cycles)	42
Graphene with sulfonic cellulose/GF	260 μm/—	2.45 × 10^{-8} S cm^{-1}	1400 h/10 mA cm^{-2} 10 mA h cm^{-2}	Zn||CNT-MnO$_2$ (95% capacity retention at 1 A g^{-1} after 1900 cycles)	49
GO/cellulose	—	—	500 h/10 mA cm^{-2} 1 mA h cm^{-2}	Zn||VO$_2$ (74% capacity retention at 10 A g^{-1} after 1000 cycles)	54
Ti$_3$C$_2$T$_x$ MXene/GF	>200 μm/3.3 N	14.4 S cm^{-1}	1180 h/1 mA cm^{-2} 1 mA h cm^{-2}	Zn||KVOH (77.9% capacity retention at 5 A g^{-1} after 1000 cycles)	53
MXene@NiO/cellulose	>200 μm/—	—	500 h/10 mA cm^{-2} 10 mA h cm^{-2}	Zn||NS/MXene@MnO$_2$ (99.2% capacity retention at 1 A g^{-1} after 200 cycles)	55
UiO-66/GF	—	20.97 S cm^{-1}	1650 h/2 mA cm^{-2} 1 mA h cm^{-2}	Zn||MnO$_2$ (85% capacity retention at 1 A g^{-1} after 1100 cycles)	72
MOF-808 and rGO/PP/PE	>200 μm/—	—	550 h/0.5 mA cm^{-2} 0.25 mA h cm^{-2}	Zn||MnO$_2$ (~100% capacity retention at 4 A g^{-1} after 2000 cycles)	73
PBPT/GF	~299 μm/4.25 MPa	11.78 S cm^{-1}	1540 h/0.5 mA cm^{-2} 0.5 mA h cm^{-2}	Zn||MnO$_2$ (86.1% capacity retention at 1 A g^{-1} after 1000 cycles)	52
Supramolecules/GF	—	—	2000 h/1 mA cm^{-2} 1 mA h cm^{-2}	Zn||MnO$_2$ (64.05% capacity retention at 1 A g^{-1} after 1000 cycles)	48
OH-terminated SiO$_2$/cellulose	—	2.06 S cm^{-1}	1000 h/5 mA cm^{-2} 1 mA h cm^{-2}	Zn||NVO (74.9% capacity retention at 1 A g^{-1} after 2000 cycles)	50
IS/SiO$_2$–OH/GF	—	10.37 S cm^{-1}	3400 h/1 mA cm^{-2} 1 mA h cm^{-2}	Zn||NVO (70.7% capacity retention at 2 A g^{-1} after 1500 cycles)	78

BTO/GF	—	—	1600 h/10 mA cm^{-2} 0.25 mA h cm^{-2}	Zn||MnO$_2$ (85% capacity retention at 1 A g^{-1} after 1800 cycles)	44
Sn/cellulose	—	—	4500 h/1 mA cm^{-2} 1 mA h cm^{-2}	Zn||MnO$_2$ (~200 mA h g^{-1} at 0.3 A g^{-1} after 600 cycles)	51
Zn-Nafion separator	25 μm/~15.1 MPa	—	553 h/5 mA cm^{-2} 0.5 mA h cm^{-2}	Zn||V$_2$O$_5$ (96.8% capacity retention at 1 A g^{-1} after 250 cycles)	86
Zn^{2+}-integrated Nafion separator	—	2.6 S cm^{-1}	—	Zn||V$_2$O$_5$ (84% capacity retention at 5 A g^{-1} after 1300 cycles)	88
ZPSAM separator	23 μm/38–41 MPa	~1.17 × 10^{-3} S cm^{-1}	2000 h/0.5 mA cm^{-2} 0.5 mA h cm^{-2}	Zn||VS$_2$ (147 mA h g^{-1} at 0.2 A g^{-1} after 100 cycles)	89
CF separator	140 μm/29.2 MPa	56.95 S cm^{-1}	2000 h/1 mA cm^{-2} 1 mA h cm^{-2}	Zn||LPC/α-MnO$_2$ (87.7% capacity retention at 1 A g^{-1} after 1000 cycles)	93
Bamboo cellulose membrane separator	~10 μm/81 MPa	—	5000 h/0.5 mA cm^{-2} 0.25 mA h cm^{-2}	Zn||MnO$_2$ (58 mA h g^{-1} at 1 A g^{-1} after 100 cycles)	98
BCM separator	9 μm/120 MPa	0.061 S cm^{-1}	4000 h/0.5 mA cm^{-2} 0.1 mA h cm^{-2}	Zn||MnO$_2$ (171.7 mA h g^{-1} at 1 C after 100 cycles)	99
CNF-SO$_3$Zn separator	25 μm/210.3 MPa	7.3 × 10^{-5} S cm^{-1}	500 h/1 mA cm^{-2} 0.5 mA h cm^{-2}	Zn||PANI (95% capacity retention at 0.2 A g^{-1} after 150 cycles)	45
ZrO$_2$ cellulose separators	50 μm/—	4.59 S cm^{-1}	2000 h/0.5 mA cm^{-2} 0.25 mA h cm^{-2}	Zn||MnO$_2$/Graphite (87.2% capacity retention at 2.5 A g^{-1} after 3000 cycles)	100
PAN separator	69 μm/3.6187 MPa	0.45 × 10^{-2} S cm^{-1}	800 h/0.283 mA cm^{-2} —	Zn||NH$_4$V$_4$O$_{10}$ (84.3% capacity retention at 5 A g^{-1} after 1000 cycles)	101
CT separators	415 μm/~6 MPa	0.01 S cm^{-1}	1200 h/1 mA cm^{-2} 0.5 mA h cm^{-2}	Zn||MnO$_2$/CNT (96.9 mA h g^{-1} at 1 A g^{-1} after 2400 cycles)	103
Filter paper separator	131 μm/16.8 MPa	4 S cm^{-1}	2600 h/1 mA cm^{-2} 1 mA h cm^{-2}	Zn||NaV$_3$O$_8$·1.5H$_2$O (83.8% capacity retention at 5 A g^{-1} after 5000 cycles)	43

the effect on the separator should be strengthened through experimental and theoretical calculations, including solvent components, zinc salt and concentrations of electrolyte.

Acknowledgments

This work was supported by the National Natural Science Foundation of China (22005268, 51972286, 21905246), the Natural Science Foundation of Zhejiang Provincial Natural Science Foundation (LQ20B010011, LR19E020003 and LZ21E020003), X. C. and F. W. also thank the support from Leading Innovative and Entrepreneur Team Introduction Program of Zhejiang (2020R01002). F. D. and F. W. contributed equally to this work.

References

1. D. Zhao et al., *Nano Energy* **72**, 104715 (2020).
2. W. Liu et al., *Nano Res.* **16**, 2325 (2023).
3. S. Gao et al., *Small Struct.* **3**, 2200086 (2022).
4. N. Ma et al., *Funct. Mater. Lett.* **12**, 1930003 (2019).
5. M. K. Aslam et al., *Mater. Today Energy* **31**, 101196 (2023).
6. S. Zhang et al., *Nano Res. Energy* **1**, e9120001 (2022).
7. Y. Liu et al., *Joule* **5**, 2845 (2021).
8. F. Xin et al., *Electrochem. Energy Rev.* **3**, 643 (2020).
9. H. Shang et al., *Angew. Chem. Int. Ed.* **57**, 774 (2018).
10. Y. Shang et al., *Curr. Opin. Electrochem.* **33**, 100954 (2022).
11. J. Ming et al., *Mater. Sci. Eng. R Rep.* **135**, 58 (2019).
12. R. Yao et al., *Adv. Energy Mater.* **12**, 2102780 (2022).
13. D. Ma et al., *Energy Environ. Mater.* **6**, e12301 (2023).
14. H. Ge et al., *Nano Res. Energy* **2**, e9120039 (2023).
15. S. Gheytani et al., *Adv. Sci. (Weinh.)* **4**, 1700465 (2017).
16. J. Niu et al., *Adv. Energy Mater.* **10**, 2000697 (2020).
17. G. A. Elia et al., *J. Power Sources* **481**, 228870 (2021).
18. Q. Wei et al., *Chem. Sci.* **13**, 5797 (2022).
19. M. Kotobuki et al., *Energy Storage Mater.* **54**, 227 (2023).
20. H. Liu et al., *J. Energy Chem.* **77**, 642 (2023).
21. J. Zheng et al., *ACS Energy Lett.* **7**, 197 (2022).

22. H. Yu *et al.*, *ACS Nano* **16**, 9736 (2022).
23. Q. Liu *et al.*, *Adv. Energy Mater.* **12**, 2200318 (2022).
24. X. Zhang *et al.*, *InfoMat* **4**, e12306 (2022).
25. J. Yang *et al.*, *Nano-Micro Lett.* **14**, 42 (2022).
26. H. He *et al.*, *Energy Storage Mater.* **43**, 317 (2021).
27. K. F. Ouyang *et al.*, *Adv. Funct. Mater.* **32**, 2109749 (2022).
28. R. Xue *et al.*, *J. Mater. Chem. A* **10**, 10043 (2022).
29. X. Zeng *et al.*, *Adv. Mater.* **33**, 2007416 (2021).
30. Y. Yang *et al.*, *Adv. Mater.* **33**, 2007388 (2021).
31. Y. Zhang *et al.*, *Chem. Eng. J.* **448**, 137653 (2022).
32. X. Liu *et al.*, *Adv. Sci. (Weinh.)* **7**, 2002173 (2020).
33. Z. M. Zhao *et al.*, *Energy Environ. Sci.* **12**, 1938 (2019).
34. L. Ma *et al.*, *Adv. Mater.* **33**, 2007406 (2021).
35. Y. Zeng *et al.*, *Adv. Mater.* **31**, 1903675 (2019).
36. C. P. Li *et al.*, *Chem. Eng. J.* **379**, 122248 (2020).
37. F. Wu *et al.*, *Small* **18**, 2202363 (2022).
38. Y. Shang *et al.*, *Adv. Funct. Mater.* **32**, 2200606 (2022).
39. J. Zhu *et al.*, *Small* **18**, 2202509 (2022).
40. M. F. Lagadec *et al.*, *Nat. Energy* **4**, 16 (2018).
41. J. Chen *et al.*, *Energy Rev.* **1**, 100005 (2022).
42. C. Li *et al.*, *Adv. Mater.* **32**, 2003425 (2020).
43. Y. Qin *et al.*, *Small* **16**, 2003106 (2020).
44. Y. C. Liang *et al.*, *Adv. Funct. Mater.* **32**, 2112936 (2022).
45. X. S. Ge *et al.*, *Adv. Funct. Mater.* **32**, 2200429 (2022).
46. L. S. Wu *et al.*, *J. Mater. Chem. A* **9**, 27408 (2021).
47. X. P. Yang *et al.*, *Chem. Eng. J.* **450**, 137902 (2022).
48. T. C. Liu *et al.*, *Energy Storage Mater.* **45**, 1074 (2022).
49. X. Zhang *et al.*, *Adv. Mater.* **34**, 2205175 (2022).
50. Y. Yang *et al.*, *ACS Appl. Mater. Interfaces* **14**, 37759 (2022).
51. Z. Hou *et al.*, *Nat. Commun.* **12**, 3083 (2021).
52. X. X. Liu *et al.*, *Chem. Eng. J.* **437**, 135409 (2022).
53. Y. Su *et al.*, *Adv. Funct. Mater.* **32**, 2204306 (2022).
54. Y. Z. Luo *et al.*, *ACS Appl. Energy Mater.* **4**, 14599 (2021).
55. Y. An *et al.*, *ACS Nano* **16**, 6755 (2022).
56. S. Bi *et al.*, *2D Mater.* **9**, 042001 (2022).
57. Z. Cao *et al.*, *Energy Environ. Mater.* **5**, 45 (2022).
58. W. Xu *et al.*, *Batteries* **8**, 293 (2022).
59. C. Shen *et al.*, *ACS Appl. Mater. Interfaces* **10**, 25446 (2018).

60. D. Xiong *et al.*, *Small* **14**, 1703419 (2018).
61. M. Naguib *et al.*, *Adv. Mater.* **23**, 4248 (2011).
62. Y. Dong *et al.*, *Adv. Funct. Mater.* **30**, 2000706 (2020).
63. Y. Chen *et al.*, *Funct. Mater. Lett.* **14**, 2130011 (2021).
64. M. S. Javed *et al.*, *Small* **18**, 2201989 (2022).
65. J. Zhang *et al.*, *Adv. Mater.* **32**, 2001093 (2020).
66. O. M. Yaghi *et al.*, *J. Am. Chem. Soc.* **117**, 10401 (1995).
67. T. Zhao *et al.*, *Coord. Chem. Rev.* **468**, 214642 (2022).
68. Q. Zhang *et al.*, *Small* **18**, 2203583 (2022).
69. S. Yuan *et al.*, *Adv. Mater.* **30**, 1704303 (2018).
70. J. M. Taylor *et al.*, *Chem. Mater.* **27**, 2286 (2015).
71. N. Maeboonruan *et al.*, *J. Sci., Adv. Mater. Devices* **7**, 100467 (2022).
72. Y. Song *et al.*, *Nano-Micro Lett.* **14**, 218 (2022).
73. Z. Wang *et al.*, *Nano-Micro Lett.* **13**, 73 (2021).
74. T. G. Hsu *et al.*, *Nat. Commun.* **14**, 225 (2023).
75. M. Mandal *et al.*, *J. Mater. Chem. A* **8**, 17568 (2020).
76. X. Wang *et al.*, *J. Membr. Sci.* **587**, 117135 (2019).
77. J. Wang *et al.*, *Nat. Energy* **4**, 392 (2019).
78. H. Gan *et al.*, *Energy Storage Mater.* **55**, 264 (2023).
79. D. Han *et al.*, *Small* **16**, 2001736 (2020).
80. Q. Lu *et al.*, *ACS Appl. Mater. Interfaces* **13**, 16869 (2021).
81. P. Xiao *et al.*, *Energy Environ. Sci.* **15**, 1638 (2022).
82. Y. C. Liang *et al.*, *Chem. Eng. J.* **425**, 131862 (2021).
83. Y. Guo *et al.*, *Energy Storage Mater.* **50**, 580 (2022).
84. F. Zhang *et al.*, *Mater. Today Energy* **17**, 100443 (2020).
85. C. Peng *et al.*, *Nano Energy* **98**, 107329 (2022).
86. B. K. Wu *et al.*, *J. Mater. Chem. A* **9**, 4734 (2021).
87. D. Yuan *et al.*, *ChemSusChem* **12**, 4889 (2019).
88. M. Ghosh *et al.*, *Energy Technol.* **7**, 1900442 (2019).
89. Y. H. Cui *et al.*, *Energy Storage Mater.* **27**, 1 (2020).
90. H. Glatz *et al.*, *ACS Appl. Energy Mater.* **2**, 1288 (2019).
91. T. D. Nguyen *et al.*, *Adv. Funct. Mater.* **29**, 1904639 (2019).
92. J. Cao *et al.*, *Adv. Energy Mater.* **11**, 2101299 (2021).
93. W. Zhou *et al.*, *Energy Storage Mater.* **44**, 57 (2022).
94. T.-W. Zhang *et al.*, *Compos. Commun.* **14**, 7 (2019).
95. H. Gwon *et al.*, *Proc. Natl. Acad. Sci. USA* **116**, 19288 (2019).
96. Z. Li *et al.*, *Adv. Mater.* **32**, 1906308 (2020).

97. Z. Li *et al.*, *Nat. Sustain.* **5**, 235 (2021).
98. J. Z. Fu *et al.*, *Energy Storage Mater.* **48**, 191 (2022).
99. Y. Zhang *et al.*, *Cell Rep. Phys. Sci.* **3**, 100824 (2022).
100. J. Cao *et al.*, *Nano Energy* **89**, 106322 (2021).
101. Y. Fang *et al.*, *Adv. Funct. Mater.* **32**, 2109671 (2022).
102. J. Zhu *et al.*, *Adv. Mater.* **34**, 2207209 (2022).
103. P. H. Cao *et al.*, *ACS Sustain. Chem. Eng.* **10**, 8350 (2022).

Chapter 3

Effect of sintering temperature on phase transformation and energy storage properties of $0.95NaNbO_3$–$0.05Bi(Zn_{0.5}Zr_{0.5})O_3$ ceramics

Ying Ge[*], Dong Liu[*], Haifeng Zhang[*], Shuhao Yan[*], Bo Shi[*] and Junjie Hao[*,†,‡]

[*]*Institute for Advanced Materials and Technology*
University of Science and Technology Beijing
Beijing 100083, P. R. China
[†]*ustbhaojunjie@126.com*

Dielectric materials with excellent energy storage performance are crucial to the development of renewable energy. In this work, we prepared $0.95NaNbO_3$–$0.05Bi(Zn_{0.5}Zr_{0.5})O_3$ (0.95NN–0.05BZZ) ceramics using solid state sintering and investigated the effect of sintering temperature on phase structure. We find that the phase transformation of ceramics occurs with the change of sintering temperature. The cubic phase (Pm-$3m$) and the antiferroelectric phase ($Pbma$) coexist at 1150°C, the ferroelectric phase ($P2_1ma$) appears at 1200°C and its phase proportion decreases with the sintering temperature increasing from 1200°C to 1280°C. Finally, we achieve the high recoverable energy storage density (W_{rec}) 0.74 J/cm^3 and efficiency (η) 71% (at 140 kV/cm) at 1150°C.

Keywords: $NaNbO_3$ ceramics; sintering temperature; phase transformation; energy storage performance.

Based on the increasing of energy demand and ongoing support of the dual carbon goal, the development of renewable green energy have drawn considerable attention. The solution to efficiently store

---

[‡]Corresponding authors.
To cite this article, please refer to its earlier version published in the Functional Materials Letters, Volume 16(8), 2350015 (2023), DOI: 10.1142/S1793604723500157.

and convert renewable energy has become an important topic. Compared with batteries, supercapacitors and solid oxide fuel cells, dielectric capacitors are widely used in pulsed systems and modern electronic devices, owing to high power density, excellent conversion efficiency and outstanding fatigue resistance.[1,2] Antiferroelectric (AFE) ceramics have a series of merits in existing energy storage materials, such as high saturated polarization (P_{max}), near-zero residual polarization (P_r) and high breakdown voltage (E_b).[3,4] Specially, sodium niobate ($NaNbO_3$) as the most typical lead-free piezoelectric ceramics, has more significant value in energy storage filed, because it is low cost and without protective atmosphere for sintering.[5] However, the similar free energy of the P phase (AFE) and Q phase (FE) that lead to their coexistence at room temperature,[4,6] which causes NN ceramics usually exhibit large P_r. It indicates that phase is an important factor affecting the properties of ceramics. Nowadays, there are two methods to decrease P_r: (1) stabilizing the AFE phase by decreasing the tolerance factor (t); (2) introducing the heterovalent ion to form the polar nano-regions (PNRs). For example, Ye et al. adopted $Bi(Mg_{2/3}Nb_{1/3})NbO_3$ to partially replace NN, which strengthens the AFE phase and disrupts the long-range polar order.[7] Meanwhile, this is almost impossible to achieve high-quality single-phase NN ceramics due to the volatilization of Na oxide at high temperature. Therefore, it is urgent to study the low temperature sintering way and phase transformation in sintering process to achieve the excellent energy storage performance.

In this paper, we add Bi_2O_3, ZnO and ZrO_2 to decrease t, which can steady the AFE phase and introduce smaller domains to reduce P_r.[8] Meanwhile, Bi_2O_3 and ZnO can be used as sintering aids to decrease the sintering temperature and the volatilization of Na oxide in NN ceramics. Based on the above analysis and investigation, we sintered 0.95NN–0.05BZZ at different temperature and analyzed the effect of sintering temperature on the energy storage properties and phase structure of ceramics.

Using Na_2CO_3, Nb_2O_5, Bi_2O_3, ZnO and ZrO_2 as raw materials, 0.95NN–0.05BZZ ceramics were prepared by traditional solid state sintering method. The materials mentioned above were weighed separately in accordance with the molar ratio. Then ball milled the powder for 4 h, and the milling speed was 300 r/min. Obtained powder was sieved by a 100 mesh screen and dried for 6 h. Subsequently, the dried powder was pre-sintered at 950°C for 5 h to obtain the main crystal phase. At this time, the powder need to be ball milled twice. The ball milling parameters were as above, but the ball milling time was 20 h. Then, the powder with 5 wt.% polyvinyl alcohol (PVA) was pressed and molded by a hydraulic press to obtain samples. The samples were degummed at 650°C for 5 h. Finally, the samples were sintered at 1150°C, 1200°C, 1250°C and 1280°C for 2 h. The physical phase and structure of the material were determined using an X-ray diffractometer (Nihon Rigaku Smartlab 9 KW). Structural Rietveld refinement analysis of the material was performed using GSAS software. Raman spectra were obtained by a Raman spectrometer (Horiba). Impedance analyzer (E4990A 20 Hz–50 MHz) was used for dielectric property testing. Energy storage performance testing use a ferroelectric analyzer (Model 609E-6).

It can be seen that the ceramic surface is dense, without obvious holes, and that the ceramic grains are irregularly polygonal and randomly oriented. The grain size distribution of the ceramics shows a normal distribution and increases from 0.939 μm to 1.214 μm as the sintering temperature increases from 1150°C to 1280°C (Figs. 1(a)–1(d)). This is due to the fact that Bi_2O_3 acts as a sintering aid and promotes the formation of the liquid phase at high temperatures. The liquid phase wets the solid particles during the sintering process and increases the surface tension, which leads to grain growth.

With the increase of sintering temperature, 0.95NN–0.05BZZ ceramics show a single structure without the impurity phase (Fig. 2(a)). The fact indicated that the compound of 0.05Bi

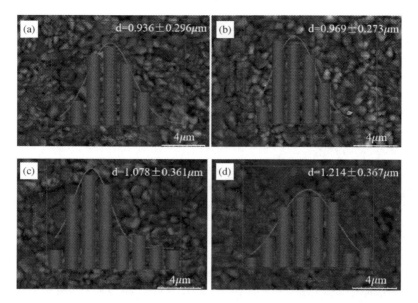

Fig. 1. The SEM image and grain distribution (a)–(d) from 1150°C to 1280°C of 0.95NN–0.05BZZ ceramics.

(Zn$_{0.5}$Zr$_{0.5}$)O$_3$ completely entered the NN lattice and formed a single and stable solid solution. Interestingly, when the sintering temperature rises from 1150°C to 1200°C, the diffraction peak starts an obvious split in the 2θ range from 56° to 60°. Meanwhile, the phase structure of ceramics remains unchanged in the temperature range of 1200°C to 1280°C (Fig. 2(a)). In order to further investigate the structure information, the XRD patterns were refined by Rietveld refinement technique (Figs. 3(a)–3(d)). The reliability factor of weighted patterns (R_{wp}), the reliability factor of patterns (R_p) and the goodness-of-fit indicator (χ^2) are found to be within the standard range, respectively. It indicates that the selected model is consistent with the actual crystal structure.

When sintering temperature is 1150°C, the *Pbma* (AFE, PDF#19-1221) and the cubic phase (*Pm-3m*, PDF#33-1270) coexist at room temperature (Fig. 3(a)). As the temperature rising, the *P2$_1$ma* (FE) replaces the pseudo-cubic *Pm-3m* paraelectrics phase. As we all known, the phase structure of NN change from *Pbcm*

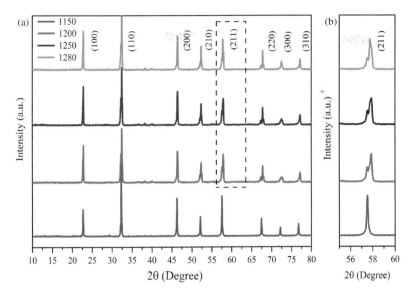

Fig. 2. XRD patterns from 1150°C to 1280°C of 0.95NN–0.05BZZ ceramics (a) 10–80° and (b) 55–60°.

phase to $Pm\text{-}3m$ phase at high temperature. Interestingly, Zhang teams reported that due to the substantial doping of Bi ions, the pseudo-cubic structure also appeared at low sintering temperature.[9] Thus, we assume the lower sintering temperature and appropriate doping ions lead to the remains of cubic phase at room temperature. The percentage of AFE phase increases from 56.3% to 73.8% in the temperature range of 1200–1280°C (Figs. 3(a)–3(d)), which could be attributed to the grain size effect.[10]

The Raman spectra of 0.95NN–0.05BZZ ceramics in the wavelength range of 50–1000 cm^{-1} were acquired to explore the structure of phase (Fig. 4). The samples are sintered at 1200°C as opposed to 1150°C undergo significant changes in the ν_5, ν_6 wave number range. It is attributed to the appearance of $Pm\text{-}3m$ cubic phase at 1150°C, which has low Raman spectral complexity due to its high symmetry.[9] Then the complexity of the spectrum increases because the $P21ma$ phase appears at 1200°C, which has a complex octahedral tilt system and B–O bond with large differences in bond

58 Y. Ge et al.

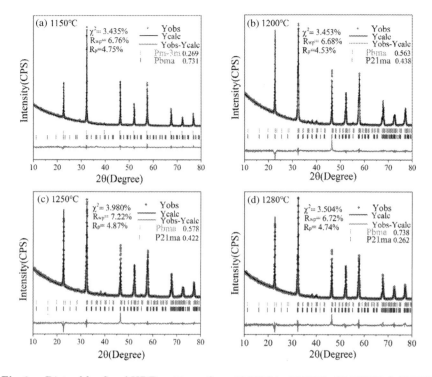

Fig. 3. Rietveld refined XRD patterns from 1150°C to 1280°C of 0.95NN–0.05BZZ ceramics (a) 1150°C, (b) 1200°C, (c) 1250°C and (d) 1280°C.

length.[11] In addition, the peak intensity decreases in the ν_4, $\nu_1+\nu_5$ wave number range, it indicates a decrease in the polarity of the basic perovskite unit cells, which favors the stabilization of the P phase.[12] With sintering temperature, there has not changed significantly, which shows the same trend as the XRD patterns.

The Curie point temperature of pure NN is around 360°C,[12] which corresponds to the transformation from P phase to R phase. The Curie point temperature gradually decreases as the sintering temperature rises from 1150°C to 1280°C (Fig. 5(d)). This is attributed to the variation of grain size following the rise in temperature. The growing concentration of P phase with the increase of grain size leads to the advance of Curie point.[10] It is also observed the dielectric peak rises at room temperature (Figs. 5(a)–5(c)). This results

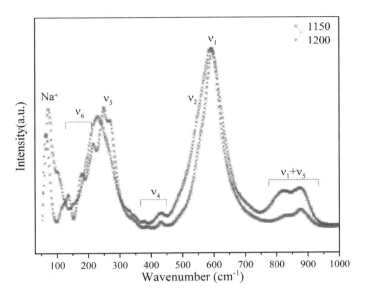

Fig. 4. The Raman spectrum at 1150°C and 1200°C of 0.95NN–0.05BZZ ceramics.

from P phase (AFE) to R phase (AFE) transformation at high temperature and R phase (AFE) to N phase (FE) transformation at low temperature move to room temperature at the same time.[12]

At the lower sintering temperature, the hysteresis loop exhibits typical AFE phase characteristics (Fig. 6(a)), which correspond to the XRD results. Interestingly, this characteristic reappears at 1280°C (Fig. 6(c)). This is because that the AFE phase influences hysteresis loop's form. The proportion of AFE phase increases at 1280°C in Raman results. It can be found that the W_{rec} shows the same trend (Fig. 6(d)), which is also attributed to the change in the proportion of the AFE phase.

We investigated the effect of sintering temperature on the phase transformation and energy storage properties of 0.95NN–0.05BZZ ceramics. With the increase of the sintering temperature, the NN ceramics underwent a phase transformation. The appropriate sintering temperature and ion doping cause the appearance of the pseudo-cubic Pm-$3m$ paraelectrics phase at 1150°C, which is beneficial to

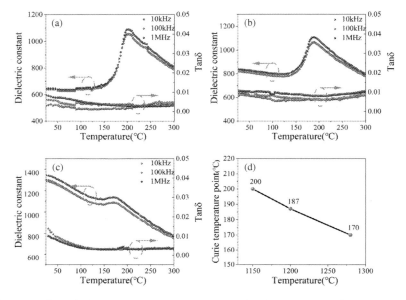

Fig. 5. The dielectric constant varies with frequency and temperature (a)–(c) from 1150°C to 1280°C and Curie point temperature variation (d) of 0.95NN–0.05BZZ ceramics.

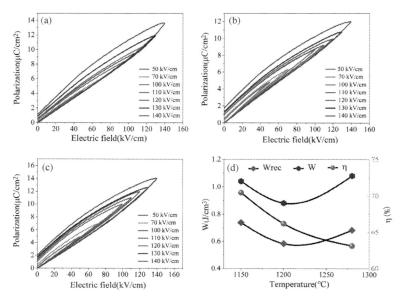

Fig. 6. The P–E loops (a)–(c) and W, W_{rec} and η (d) from 1150°C to 1280°C of 0.95NN–0.05BZZ ceramics.

obtain higher energy storage performance. When sintering temperature above 1200°C, the *Pbma* phase and *P21ma* phase dominated. Meanwhile, the phase proportion of *Pbma* and *P21ma* will change owing to the grain size at various sintering temperature. We found that the way to prepare NN ceramics at low temperature, and sintering temperature affects the phase structure of NN ceramics.

Acknowledgments

This research did not receive any specific grant from funding agencies in the public, commercial, or not-for-profit sectors.

References

1. B. Andrew, *Appl. Sci.* **11**, 8063 (2021).
2. H. Ogihara, C. A. Randall and S. Trolier-McKinstry, *J. Am. Chem. Soc.* **92**, 110 (2009).
3. F. Hua *et al.*, *ACS Appl. Mater.* **14**, 17987 (2022).
4. H. Guo *et al.*, *J. Appl. Phys.* **117**, 214103 (2015).
5. D. Yang *et al.*, *J. Mater. Chem. A* **8**, 23724 (2020).
6. H. Shimizu *et al.*, *Dalton Trans.* **44**, 10763 (2015).
7. G. Ye *et al.*, *J. Mater. Chem. C* **7**, 5639 (2019).
8. N. Qu, H. Du and X. Hao, *J. Mater. Chem. C* **7**, 7993 (2019).
9. L. Zhang *et al.*, *J. Mater. Chem. C* **9**, 4289 (2021).
10. J. Koruza *et al.*, *Acta Mater.* **126**, 77 (2017).
11. Y. I. Yuzyuk *et al.*, *J. Phys. Condens. Matter* **17**, 33 (2005).
12. J. Chen, H. Qi and R. Zuo, *ACS Appl. Mater. Interfaces* **12**, 32871 (2020).

Chapter 4

Heteroatom (N, P)-driven carbon nanomaterials for high-energy storage in supercapacitors

Yong Liu*, Xian Zhang*, Ke Zhang*, Zifeng Wang* and Guang Li[†,‡]

*School of Electronic Engineering, Huainan Normal University
Huainan 232038, P. R. China
[†]School of Materials Science and Engineering, Anhui Key Laboratory of Information Materials and Devices Key Laboratory of Structure and Functional Regulation of Hybrid Materials of Ministry of Education Institute of Physical Science and Information Technology, Anhui University
Hefei 230601, P. R. China
[‡]liguang1971@ahu.edu.cn

Nanoflower-like hollow carbon sphere (CS) materials with two different heteroatoms doping (N, P) were prepared by a simple synthetic method (NPCS). The specific surface area of NPCS can reach a high value of 396 m^2/g. The NPCS has a high degree of hollowness and the self-assembled nanosheet of NPCS forms a fast electron transport channel, and also increases active area in contact with electrolyte. The doping of heteroatoms increases the defect level of the carbon-based nanomaterials and changes the local electron state of the material, thus forming storage sites on the surface of the material, which can be used as a station for ions collecting and distributing. The material was studied as an active material for supercapacitors, and the specific capacitance reached 274.9 F/g. After the 4000th cycle stability test, it still maintained 95.2% of the specific capacitance, indicating that the material has excellent properties of supercapacitor materials.

Keywords: Carbon materials; heteroatomic doping; energy storage and conversion; nanocomposites; supercapacitor.

[‡]Corresponding authors.
To cite this article, please refer to its earlier version published in the Functional Materials Letters, Volume 16(8), 2351010 (2023), DOI: 10.1142/S1793604723510104.

1. Introduction

In recent years, with the rapid growth of the world's population and the rapid development of human civilization, electric power has become an indispensable main source of power in people's lives.[1,3] It shows extreme dependence on electric power in life entertainment, transportation, and industrial production.[4,6] The growth of electricity demand has led to the continuous growth of energy demand, but the rapid depletion of natural resources such as fossil fuels and rapid growth of pollutants and carbon emissions are also important motivations for many researchers around the world to seek environment-friendly or low-pollution alternatives and renewable energy.[7,9] In recent years, as an electrochemical energy storage device, supercapacitor has become a research hotspot due to their large capacity, long life, and wide temperature range,[10,12] and supercapacitors have been widely used in many fields. Supercapacitors are common surface/interface electrochemical energy storage devices, whose energy storage mainly comes from the electrode surface and near surface, as well as the electrode/electrolyte interface, which can achieve rapid charging and discharging.[13,16] The electrodes are the main components of supercapacitors, and the surface roughness and specific surface area of electrode materials can fundamentally impact the energy storage performance of supercapacitors.[17-19] Recently, due to its excellent electrical conductivity, controllable surface roughness, pore size, and hollow morphological characteristics, carbon material has become a very promising inexpensive electrode material.[20] The doping of heteroatoms makes the physical and chemical properties of carbon materials more suitable for the application of catalyst materials and storage materials.[21-23]

Here, we report a hollow flower-like carbon sphere (CS) material which is doped with two strongly electronegative heteroatoms (N, P) as an active material (NPCS) for supercapacitors. The doping of N and P makes the hollow sphere of carbon nanosheets form certain defects and changes their local electronic state, thereby forming an excellent electrode material with multiple active sites, large specific surface area and high degree of hollowness.[24,25]

The specific capacitance of the supercapacitor using NPCS as the electrode active material reaches 274.9 F/g, and the value of single heteroatom doped NCS, PCS, or undoped CS is less than 253 F/g. In the long-term stability experiment, NPCS also has an absolute advantage among the four subjects, and 4000 times at 1 A/g of charge and discharge experiments can still maintain 95.2% of the specific capacitance.

2. Experimental

All chemicals used in this study were analytical grade and used without further purification.

2.1. *Synthesis of CS, NCS, PCS, and NPCS*

The preparation steps are shown in Fig. 1 (①–③), all synthetic steps were performed at room temperature. Put 20 ml of deionized water (H_2O), 40 ml of absolute ethyl alcohol (C_2H_5OH), and 2 ml of ammonia water ($NH_3 \cdot H_2O$) into a 100 ml beaker, and stir for 30 min to mix well. Subsequently, 0.8 g of phenol (C_6H_6O) was added to the mixed solution and stirred for 15 min, and then 1 ml of formaldehyde solution (CH_2O) was added to continue stirring for 15 min. Finally, 5 ml of tetrapropoxy silane ($C_{12}H_{28}O_4Si$) was added to the above mixture, and then the solution was transferred to a constant temperature heating and stirring table, kept at 50°C and vigorously stirred for 18 h. The mixed solution was washed with alcohol and deionized water to obtain the precursor. The precursor was placed in a tube furnace and annealed at 850°C for two hours under N_2 atmosphere. After cooling to room temperature, the black powder was poured into 2M sodium hydroxide solution (NaOH) and kept at 45°C for 24 h. Then, the mixture was washed with deionized water to neutrality to obtain CS.

The preparation steps are shown in Fig. 1 (①–④), all synthetic steps were performed at room temperature. 100 mg of CS was mixed with 1 g of melamine ($C_3H_6N_6$), the mixture was put into a

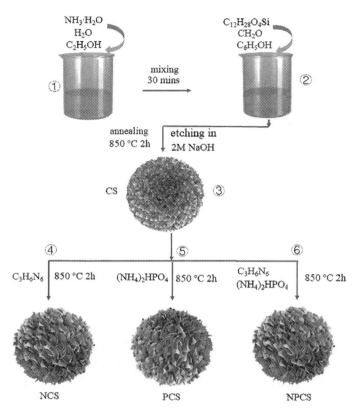

Fig. 1. CS, NCS, PCS, NPCS preparation process.

quartz boat, and transferred to an annealing furnace, annealed at 850°C under N_2 protection, kept for 2 h, and cooled to room temperature to form NCS.

The preparation steps are shown in Fig. 1 (①–③, ⑤), all synthetic steps were performed at room temperature. 0.5 g ammonium phosphate dibasic $(NH_4)_2HPO_4$ was placed in an alumina crucible, upstream of the tubular annealing furnace, 100 mg CS was placed in another alumina crucible, placed downstream, N_2 was passed from the upstream, annealed at 850°C for 2 h, and cooled to room temperature form PCS.

The preparation steps are shown in Fig. 1 (①–③, ⑥), all synthetic steps were performed at room temperature. 0.5 g ammonium phosphate dibasic $(NH_4)_2HPO_4$ was placed in an alumina crucible, placed upstream of the tubular annealing furnace, 100 mg of CS was mixed with 1 g melamine $(C_3H_6N_6)$ and placed in another alumina crucible, placed downstream, and N_2 was passed from upstream, annealed at 850°C 2 h, cooled to room temperature to form NPCS.

2.2. Preparation of working electrodes

The foam nickel was into 2 × 1 cm rectangle, and clean it with alcohol ultrasonic to ensure full cleaning. Vacuum dry foam nickel sheets to remove alcohol. During drying, isolate air to avoid oxidation of foam nickel. Then bend the rectangular foam nickel along the middle to form an *L* shape. Then, prepare a blend solution of PTFT: anhydrous ethanol: carbon black: material = 1:1:1:8. After the sample solution ratio is completed, grind evenly to form a colloid. Then use a clean small brush to evenly coat the mixed solution on the inner surface of *L*-shaped foam nickel, and then vacuum dry it tills completely dry. Then, fold it along the inner surface to form a 1 × 1 cm double-layer foam nickel. The material is attached to the interlayer and pressed into thin sheets under the pressure of 20 MP, so that the working electrode is ready.

3. Results and Discussions

3.1. Morphology analysis

Observing the surface and internal morphology of materials by electron microscopy. Figure 2(a) is the scanning electron microscopy (SEM) image of NPCS at 500 nm scale. It can be seen that the carbon nanosheets grow in an orderly manner, forming a flower shape. Figure 2(b) is SEM image of NPCS at 300 nm scale, there is a broken hollow sphere in the image, which shows the hollow

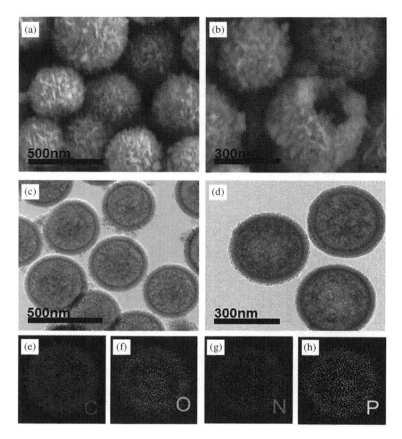

Fig. 2. (a), (b) SEM of NPCS, (c), (d) TEM of NPCS, (e), (f) EDS mapping of C, O, N, and P species.

characteristics of the material. Figures 2(c) and 2(d) are the transmission electron microscopy (TEM) of NPCS at 500 nm and 300 nm scale, respectively. From the image, it can be seen that the spheres have basically the same size, with a diameter of around 400 nm, and are evenly distributed in space. It can further confirm the hollow structure of the material and the ordered growth of nanosheets around the sphere. Of particular note, the inner space (diameter = 300 nm) and outer nano nanosheet shell (thickness = 50 nm) of NPCS can provide rich electrochemical active sites, thereby obtaining higher specific capacitance.[26,27] Interconnection channels between external nanosheets can improve the electrolyte wettability of materials,[28] and thin shell walls can shorten the transport distance between ions and electrons, improve the electron transport rate, and thereby increase the reaction rate.[29-31] The energy-dispersive spectroscopy (EDS) mapping test can provide a more intuitive understanding of the composition and element distribution of materials, and Figs. 2(e)–2(h) are the distribution areas of C, O, N, and P elements, respectively, and each element is evenly distributed in the CS, indicating that N and P were successfully and uniformly doped into the CS.[32,33] The specific surface area of the material was further clarified using the N_2 adsorption/desorption analysis test, and the NPCS reached 396 m^2/g (Fig. 3(a)). So, this hollow flower structure has a large specific surface area, which can increase the contact area between material and electrolyte.

3.2. Composition analysis

X-Ray diffraction (XRD) is usually used to analyze the composition of materials. Figure 3(b) is the XRD pattern of materials. It can be seen that there are obvious broad peaks at the 23° and 44° position, which is a typical carbon peak, corresponding to (002) and (101) crystal planes, respectively, indicating that the main component of the material is carbon. Raman spectroscopy can further analyze the structure information of the material. As shown in Fig. 3(c), there are peaks at the 1337 cm^{-1} and 1581 cm^{-1} positions, respectively, representing the D and G bands. The D band indicates the disorder or defect in the graphitic structure, the G band represents the graphitic order, the intensity ratio (I_D/I_G) of the D band and the G band indicates the degree of disorder or defect in the carbon material.[34,35] The I_D/I_G ratios of CS and NPCS are 0.87 and 1.07, respectively, obviously, the defects of NPCS the degree are stronger than CS.

In order to further confirm the state in which each element in the material exists and the bond state of the elements with each other, we tested X-ray photoelectron spectroscopy (XPS). The C(1s) spectrum (Fig. 3(d)) shows that the C–P and C–N bonds are

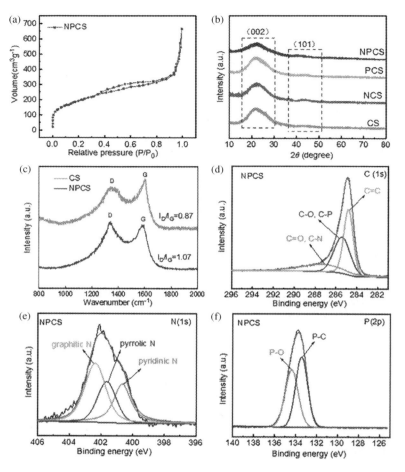

Fig. 3. (a) N$_2$ adsorption/desorption isotherms of NPCS, (b) XRD of each material, (c) Raman of each material, XPS spectra of NPCS: (d) C(1s) spectra, (e) N (1s) spectra, (f) P (2p) spectra.

located at 285 and 287 eV, respectively, it indicates that the C, O, N, P elements of NPCS are consistent with the mapping results. Figure 3(e) shows the N(1s) spectrum of NPCS, which can be divided into three peaks, corresponding to pyrrolic N, pyridinic N, and graphitic N. The P(2p) spectrum (Fig. 3(f)) can be found at 133.5 and 134.4 eV, corresponding to P–C and P–O bonds, respectively. The above analysis shows that the N and P heteroatoms are successfully and uniformly doped into the carbon lattice.[36,38]

3.3. Electrochemical analyses

The electrochemical performance of the entire samples was systematically evaluated using a three-electrode system (Ag/AgCl as the reference electrode, Pt foil as the counter electrode, and active material as the working electrode) with 6 M KOH as the electrolyte solution. Cyclic voltammetry (CV) is an important test method to evaluate the properties of active materials for supercapacitors. Figure 4(a) shows the CV curves of the active materials of CS, NCS, PCS, and NPCS in the range −1–0 V with a scan rate of 20 mV/s. Comparing the quasi-rectangular shapes of CS, the CV curves of NCS, PCS, and NPCS are not only quasi-rectangular, but also have peaks broadened due to the redox reactions of N and P containing groups, indicating the coexistence of double-layer capacitance and pseudo-capacitance.[39] From the CV curves, we can clearly understand that the integral area of NPCS is larger than the other three active materials. Figure 4(b) shows the CV curves

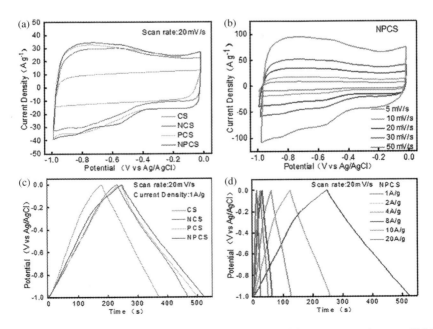

Fig. 4. (a) CV curves (20 mV/s) of each material, (b) CV curves (5–50 mV/s) of NPCS, (c) GCD (1 A/g) of each material, (d) GCD (1–20 A/g) of NPCS.

of NPCS at different scan rates from 5 mV/s to 50 mV/s. All CV curves exhibit approximately rectangular shapes with distinct peak shapes, reflecting the complete capacitance mainly from the electric double layer capacitance (EDLC) and the auxiliary pseudo-capacitance.

In order to fully analyze the energy storage capacity of the active material, we conducted a constant current charge-discharge test on the material. The test was carried out under the conditions of a scan rate of 20 mV/s and a current density of 1 A/g. It can be seen from Fig. 4(c) that due to the existence of EDLC, the curves are all isosceles triangles and due to the pseudo-capacitance effect of N and P doping, the curves are slightly curved. The storage capacity of NPCS (274.9 F/g) is ahead of the other materials, and the specific values are shown in Table 1. Rate capacity is also an indispensable advantage of a good active material. In Fig. 4(d), we can observe that the NPCS maintains satisfactory specific capacitance (247.5 F/g), even at the current density of 20 A/g, the other specific values of materials are shown in Fig. 5(a) and Table 1. Meanwhile, the cycling durability performance of the material was tested and analyzed, as shown in Fig. 5(b), NPCS showed excellent cycling durability, maintaining 95.2% of the specific capacitance after 4000 cycles. Because the hollow structure can buffer the volume expansion during charging and discharging, improve the material cycle stability, and the self-assembled nanosheet sheet hollow

Table 1. Specific capacitance C_m (different current densities) and long-term stability (cycling stability) of each material.

	C_m (F/g) (5 mV/s)						Cycling Stability	
I/g	1 A/g	2 A/g	4 A/g	8 A/g	10 A/g	20 A/g	4000th	Retention
CS	195.1	192.2	189.2	188.8	182.9	181.3	165.9	82.5%
NCS	243.9	236.0	228.5	221.7	219.0	213.1	205.7	85.1%
PCS	252.6	245.3	241.9	238.7	234.4	227.3	228.3	90.4%
NPCS	274.9	266.9	259.6	255.5	254.7	247.5	264.9	95.2%

Heteroatom (N, P)-driven carbon nanomaterials for high-energy storage 73

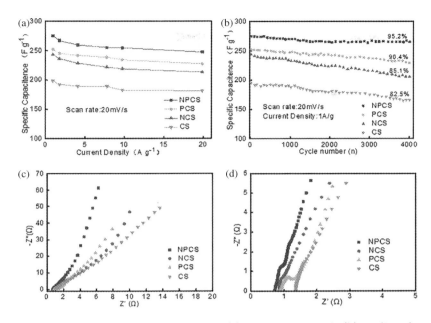

Fig. 5. (a) Specific capacitance (1–20 A/g) of each material, (b) cycling durability performance (4000th) of each material, (c), (d) Nyquist plots of each material.

structure has better mechanical properties due to the mutual support of different sheets. The conductivity of the materials determines the electron migration rate, which is related to the electrochemical performance of the materials. Figures 5(c) and 5(d) show the typical Nyquist diagram of material. The semi arc in the high frequency and the oblique line at the low frequency, respectively, illustrate the redox activity and diffusion processes at the interface. The Nyquist diagram of NPCS has smaller semicircle and steeper oblique lines, showing relatively fast charge transport dynamics and better ion diffusion ability.[40]

4. Conclusions

In this work, carbon flower balls doped with two kinds of heteroatoms were studied as active electrodes of supercapacitors. There are two main reasons for its high performance: first, the hollow structure of the material significantly increases the contact area between the active site and the electrolyte, and at the same time, self-assembled nanosheet structure will establish ion and electron transport pathways extending in all directions and improve the reaction kinetics; second, the dual doping of N and P elements changes the local electronic state of the material, resulting in more defects and enhancing the overall wettability and compatibility between the material and electrolyte, enhancing the reversible Faraday process, endowing the composite material with ultrafast charge storage kinetics and excellent electrochemical properties. The NPCS have enormous application potential in energy storage field and it can be used as an efficient electrode material for non-precious metals.

Acknowledgments

This work was financially supported by GCCRCKYQDJ (No. 621243), 2022HX66 (No. 722087), Natural Science Foundation in Universities of Anhui Province (No. KJ2019A0689), Anhui Province University Excellent Talents Support Program (No. GXYQ2022069), Key Projects of Natural Science Research in Colleges and Universities of Anhui Province rank (No. KJ2021A0975).

References

1. K. Siala *et al.*, *Renew. Energy* **88**, 445 (2016).
2. H. Herzog *et al.*, *Encyclopedia Energy* **1**, 277 (2004).
3. M. A. H. Mondal *et al.*, *Renew. Energy* **36**, 1869 (2011).
4. A. Yushchenko *et al.*, *Renew. Sustain. Energy Rev.* **81**, 2088 (2018).
5. Z. Li *et al.*, *Adv. Energy Mater.* **13**, 2203019 (2023).
6. J. Liang *et al.*, *Appl. Energy* **266**, 114851 (2020).
7. A. Rezvani *et al.*, *Front. Energy* **13**, 131 (2019).
8. S. Lal *et al.*, *Energy Sources A Recovery Util. Environ. Eff.* **42**, 1914 (2020).
9. A. Abdollahipour *et al.*, *Appl. Therm. Eng.* **187**, 116576 (2021).

10. Q. Zhu *et al.*, *Adv. Energy Mater.* **9**, 1901081 (2019).
11. P. Yu *et al.*, *Adv. Energy Mater.* **6**, 1601111 (2016).
12. K. Sadasivuni *et al.*, *J. Mater. Sci. Mater. Electron.* **30**, 951 (2019).
13. H. Dong *et al.*, *ES Energy Environ.* **4**, 19 (2019).
14. H. Wei *et al.*, *J. Alloys Compd.* **820**, 153111 (2020).
15. R. Nie *et al.*, *Eng. Sci.* **6**, 22 (2018).
16. W. Wang *et al.*, *Nanoscale Horiz.* **4**, 1211 (2019).
17. J. Hao *et al.*, *J. Colloid Interface Sci.* **537**, 57 (2019).
18. S. He *et al.*, *J. Power Sources* **467**, 228324 (2020).
19. W. Li *et al.*, *Adv. Mater.* **20**, 5129 (2013).
20. P. Zhang *et al.*, *Chem. Eur. J.* **23**, 1986 (2017).
21. X. Zhu *et al.*, *Chem. Sus. Chem.* **12**, 1 (2019).
22. K. Jayaramulu *et al.*, *Adv. Mater.* **30**, 1705789 (2018).
23. S. Ghosh *et al.*, *Adv. Energy Mater.* **10**, 2001239 (2020).
24. R. Anurag *et al.*, *J. Energy Storage* **56**, 106013 (2022).
25. F. Yang *et al.*, *Electrochim. Acta* **328**, 135064 (2019).
26. J. Wang *et al.*, *Research* **2022**, 9837012 (2022).
27. Z. Cui *et al.*, *Small Methods* **6**, 2101484 (2022).
28. M. Shang *et al.*, *Appl. Surf. Sci.* **542**, 148697 (2021).
29. K. Xu *et al.*, *Adv. Mater.* **28**, 3326 (2016).
30. L. Sun *et al.*, *Chin. J. Catal.* **40**, 80 (2019).
31. X. Luo *et al.*, *J. Mater. Chem. A* **8**, 17558 (2020).
32. K. Shang *et al.*, *J. Mater. Chem. A* **10**, 6489 (2022).
33. T. Zhao *et al.*, *Appl. Surf. Sci.* **569**, 151098 (2021).
34. A. Cuesta *et al.*, *Carbon* **32**, 1523 (1994).
35. K. Xia *et al.*, *J. Power Sources* **365**, 380 (2017).
36. Y. Zheng *et al.*, *Small* **16**, 2004342 (2020).
37. V. Duraisamy *et al.*, *Chem. Select* **6**, 11887 (2021).
38. L Li *et al.*, *ACS Sustain. Chem. Eng.* **7**, 1337 (2019).
39. Y. Yang *et al.*, *Nano-Micro Lett.* **9**, 6 (2017).

Chapter 5

Impact of annealing on charge storage capability of thermally evaporated molybdenum oxide thin films

Sudesh Kumari*, Rameez Ahmad Mir[†,§], Sanjay Upadhyay[‡], O. P. Pandey[§] and Anup Thakur*[¶]

*Department of Physics, Punjabi University Patiala
Punjab, 147002, India
[†]The University of British Columbia
3333 University Way, Kelowna, BC V1V1V7, Canada
[‡]Division of Research and Innovation, Uttaranchal University
Dehradun 248007, India
[§]School of Physics and Materials Science Thapar Institute
of Engineering and Technology, Patiala, Punjab, India
[¶]dranupthakur@gmail.com

Rapid technological advancements in recent years have necessitated the creation of energy-related devices. Due to their unique properties, including an exceptional cycling life, safe operation, low processing cost, and a higher power density than batteries, supercapacitors (SCs) have been identified as one of the most promising candidates to meet the demands of human sustainability. To increase the energy density of SCs, several advanced electrode materials and cell designs have been researched during the past few years. Utilizing the Faradaic charge storage process of transition metal cations, transition metal compounds have recently received attention as prospective electrode materials for SCs with high energy densities. In this work, the structure, morphology and electrochemical properties of molybdenum oxide (MoO_3) films deposited via thermal evaporation technique and annealed at various temperatures were systematically investigated in order to examine the

---

*Corresponding authors.
To cite this article, please refer to its earlier version published in the Functional Materials Letters, Volume 16(8), 2340018 (2023), DOI: 10.1142/S1793604723400180.

potential use of MoO$_3$ for supercapacitors. Electrochemical analysis confirmed the pseudocapacitance characteristics of the synthesized films. The annealing temperature affects the oxidation and reduction observed in the cyclic voltammetry (CV) plots. Areal capacitance of thin films annealed at 150°C was found to be maximum and this could be attributed to the formation of hollow tube-like nanostructures which provided more active sites, than films annealed at higher temperatures. This also influences the charge storage ability of the synthesized films. It would be logical to assume that additional research in this area will result in more interesting discoveries and, eventually, the vi-ability of those promising Mo-based compounds in high-tech energy storage systems.

Keywords: Molybdenum oxide; supercapacitor; thin films; cyclic voltammetry.

1. Introduction

The swift development of economy and the advances in technology have led to the invention of highly efficient and low-cost energy storage devices such as batteries and supercapacitors. Supercapacitors, also known as electrochemical capacitors or ultracapacitors, have attracted much attention due to their high power density compared to batteries, long cycle life and higher energy density compared to conventional electrostatic capacitors.[1-4] These extraordinary characteristics of supercapacitors enable their use in power source applications such as electric vehicles and electronic devices.[5] The electrochemical performance of supercapacitors can be increased by developing suitable electrode materials having such compositions and structures that enable the process of electron transfer and ion diffusion to occur rapidly and reversibly.

Based on charge/discharge mechanism, the supercapacitors are classified into two categories: electrical double layer capacitors (EDLC) and pseudocapacitors. EDLCs commonly use reversible adsorption mechanism (non-Faradaic process) of charge storage whereas pseudocapacitors use redox reaction mechanism (Faradaic process) of charge storage. Usually, carbon materials with high surface areas act as electrodes for EDLCs, which have low energy density since these store charges by adsorption of ions on material surface. A lot of research has been carried out to improve the energy density of supercapacitors. The conducting polymers and transition metal oxides (TMOs), with high energy density, are the promising candidates as electrode materials for supercapacitors, especially the pseudocapacitors. TMOs have gained research interest as electrode materials for pseudocapacitors on account of their high abundance, environment friendly, and other intriguing characteristics such as diverse constituents and morphologies, large surface area and high theoretical specific capacitance.[6] Nowadays there is widespread demand of TMO-based pseudocapacitor materials due to their higher energy density enhanced by defect engineering. Different TMOs have been developed for supercapacitor applications but the lower electrical conductivity restricts their practical use. The high pseudocapacitances of TMOs enable them to be a promising candidate for the next generation of supercapacitors.[7]

Among the various TMOs, molybdenum oxide (MoO_3) with layered structure is a very attractive material for charge storage as it can easily intercalate ions in a wide range of sites within the layered structure. It consists of corner sharing chains of MoO_6 octahedra that share edges with two similar chains to form layers of MoO_3 stoichiometry. These layers are held together by weak van der Waals forces. This layered structure provides sufficient space to ions for intercalation when subjected to electrochemical reaction. Molybdenum oxide, especially MoO_3, has several important properties such as a wide range of stoichiometry and different oxidation states and it is employed in a variety of technological applications such as solar cells,[8,9] organic light emitting devices,[10] gas and chemical sensors[11] and storage devices.[12-14] MoO_3 particles are being widely explored for a number of energy conversion and storage applications that include supercapacitors and batteries, owing to the charge storage capability of molybdenum oxide, which is entirely due to its layered structure. However, the MoO_3 thin films having better surface features are yet to be explored for energy storage applications.

Various deposition techniques have been employed to deposit molybdenum oxide thin films that include chemical vapor deposition,[15] thermal evaporation,[16] sol-gel coating,[17] sputtering,[18,19] electrodeposition,[20] flash evaporation[21] and spray pyrolysis technique.[22] Among these, thermal evaporation being a low cost, reliable and highly efficient technique is widely used for depositing thin films of molybdenum oxide. Molybdenum oxide can be easily deposited as thin films with this technique since its melting and evaporation point[23] can be easily achieved by thermal evaporation. The films deposited by thermal evaporation are usually amorphous and sub-stoichiometric MoO_{3-x}.[23,24] These sub-stoichiometric films find applications in photochromic, thermochromic and electrochromic devices.[25,26]

This work discusses the electrochemical behavior of thermally evaporated thin films of MoO_3. In this work, electrochemical properties of thermally evaporated thin films of molybdenum oxide (MoO_3) deposited on FTO glass substrates were studied by employing techniques like cyclic voltammetry (CV) and electrochemical impedance spectroscopy (EIS). The CV at multiple scan rates revealed the supercapacitive behavior of the fabricated thin films of molybdenum oxide. It was observed that upon evaporation, the morphology of these films changed leading to the formation of different nanostructures like nanorods and nanotubes which affected the oxidation and reduction occurring in these films, as a result of which the number of active sites varied, thus leading to enhanced super capacitive behavior in certain films.

2. Experimental Details

Thin films of molybdenum oxide were deposited on FTO glass substrates using thermal evaporation technique.[27,28] MoO_3 powder (99.99% pure, Sigma Aldrich) as source material has been used. High vacuum (5×10^{-6} mbar) inside the deposition chamber was maintained throughout the deposition. The rate of deposition ($\approx$ 3–4 A/sec) and thickness of films were measured *in-situ* by a digital thickness monitor (Hind High Vacuum DTM-101). Prior to the deposition of films, substrates were washed with water and cleaned ultrasonically at a temperature of 50°C for 45 min. Then the substrates were cleaned with acetone and methanol and heated at 80°C for 1 h to remove any volatile impurities. The as-deposited films were annealed in air at 150°C, 250°C and 350°C using muffle furnace.

Electrochemical properties of these films were investigated using multichannel galvanostat/potentiostat made of Biologic (VSP300). The film served as the working electrode with Pt as the counter/reference electrode calibrated with Ag/AgCl as the reference electrode. The electrolyte used was 0.5 M LiClO$_4$ dissolved in propylene carbonate. The CV measurements were in a wide voltage window −1 to +1 V for 30 cycles to confirm the oxidation/reduction behavior of the synthesized films. Cyclic voltammetry (CV) measurements were performed at various scan rates in voltage window of 0 to −0.4 V to confirm the supercapacitor behavior of the thin films.

3. Results and Discussion

3.1. *Structural study*

The XRD patterns of the as-deposited and annealed films of MoO$_3$ were recorded using X-ray diffractometer with Cu-$K\alpha$ radiation ($\lambda \sim 1.5406$ Å) as shown in Fig. 1. The XRD patterns of as-deposited thin film confirmed the amorphous nature since no sharp peak was observed in the patterns. At the annealing temperature of 150°C, the XRD pattern showed a peak that corresponded to α-MoO$_3$ phase (ICDD no. 00-005-0508). At 250°C, the observed XRD peaks corresponded to reflections of orthorhombic α-MoO$_3$ phase. At 350°C, the relatively intense peaks at lower angles corresponded to substoichiometric Mo$_5$O$_{14}$ phase (ICDD no. 01-074-1415) along with the reflections of orthorhombic phase. These peaks observed in the spectrum of annealed molybdenum oxide thin film indicated the improvement in the crystallization of the films at higher annealing temperatures. These results are reported elsewhere, in Ref. 29.

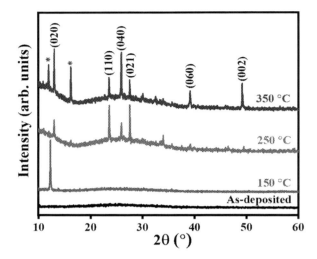

Fig. 1. X-ray spectra of as-deposited and annealed molybdenum oxide thin films.[29]

3.2. Morphological study

The morphology of the as-deposited and annealed films of MoO_3 was analyzed with FE-SEM technique (HITACHI SU8010). It was observed from Fig. 2 that the morphology of as-deposited thin film consisted of agglomerated nanoclusters.

With annealing at 150°C, these agglomerated nanoclusters changed to tube-like nanostructures having large length. With annealing at 250°C, these tube-like nanostructures changed to rod-like structures but the length of these rods has decreased. Further, these nanorods agglomerated and fused lengthwise at annealing temperature of 350°C.[29] The formation of nanotubes and nanorods goes well with coexistence of various phases and interfaces with annealing as observed from reflections of XRD spectra too. The aspect ratio (ratio of length to width) of the nanotubes formed at 150°C was 3.38 while that of nanorods formed at 250°C was 4.03. The aspect ratio of the structures formed at 350°C was 6.10.

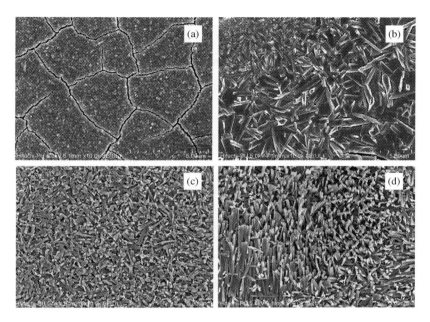

Fig. 2. FESEM images of surface morphology of MoO₃ thin films (a) as-deposited, (b) annealed at 150 °C, (c) annealed at 250°C and (d) annealed at 350°C.

3.3. *Electrochemical study*

3.3.1. *Cyclic voltammetry*

The cyclic voltammetry measurements were made to study the electrochemical properties of the as-deposited and annealed films of MoO₃. The CV performed in the wide voltage window (Fig. 3) revealed that the deposited films are active electrocatalyst species for reduction reactions due to the rapid increment in current in the reduction region. Only the as-deposited film shows less activity due to its amorphous characteristics and larger particles size. The CV pattern of the as-deposited film and the film annealed at 150°C does not reveal any specific phase changes occurring during the continuous oxidation/reduction process (CV analysis). CV patterns of the samples annealed at 250°C and 350°C exhibited the presence

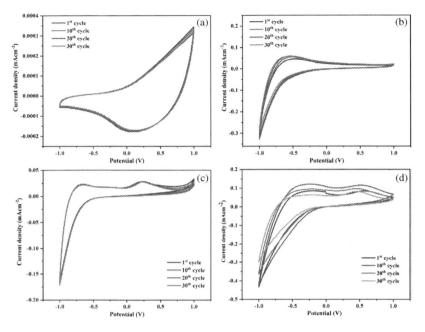

Fig. 3. Cyclic voltammograms at 50 mVs^{-1} of MoO$_3$ thin films for different cycles: (a) as-deposited, (b) annealed at 150°C, (c) annealed at 250°C and (d) annealed at 350°C.

of oxidation peaks at 0.3 V, which might be attributed to the phase changes occurring during intercalation/deintercalation of ions. The results also may be attributed to the enhancement of redox reactions occurring in the oxygen-deficient Mo-oxide phases formed due to relatively higher annealing temperatures.[30]

It is observed from cyclic voltammetry analysis that heat-treated films of molybdenum oxide exhibit sharper oxidation peaks. Upon heat treatment, the structure of the films changes from amorphous to crystalline, which being more structurally ordered might exhibit higher number of active sites for activation during CV process. This results in oxidation peaks becoming sharper in annealed films.[31] To confirm the supercapacitor behavior of the deposited thin films, the CV at multiple scan rates was performed as shown in Fig. 4.

The nearly rectangular shape of the CV pattern indicates the pseudocapacitance behavior of the synthesized films.[4,32,33] Earlier, the CV pattern exhibiting redox peaks was considered as that of a

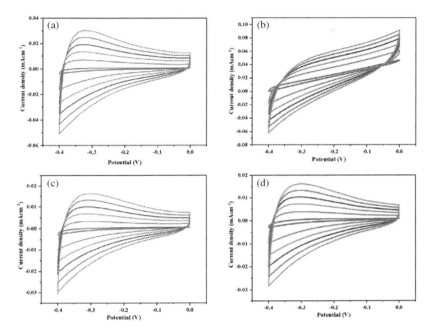

Fig. 4. Cyclic voltammograms of MoO₃ thin films at various scan rates of 5 mVs⁻¹, 50 mVs⁻¹, 100 mVs⁻¹, 150 mVs⁻¹, 200 mVs⁻¹ and 250 mVs⁻¹: (a) as-deposited, (b) annealed at 150°C, (c) annealed at 250°C and (d) annealed at 350°C.

pseudocapacitors. After that Jiang and Liu[34] and Brousse et al.[35] differentiated the confusion between battery electrode and capacitors based on the CV shapes and revealed that pseudo-capacitor also exhibits a quasi-rectangular CV shape rather than having oxidation/reduction peaks in non-Faradaic region. This attributes to the rapid transfer of ions crossing the double layer. The CV plots (Fig. 4) show that the areal capacitance increases with increasing scan rates, which confirms the enhanced charge storage capability of the synthesized films.[36]

3.3.2. *Working mechanism of MoO₃ thin films supercapacitor*

It is suggested that the capacitance of MoO₃ thin films should originate from the intercalation/deintercalation of Li⁺ as discussed in this section. During intercalation/deintercalation process, there is the possibility of occurrence of two mechanisms: the first mechanism involves the simple surface adsorption of alkali metal cations Li$^+$ in the electrolyte on the surface of the electrode material and the second mechanism involves the intercalation of alkali metal cations Li$^+$ in the electrolyte into the electrode material during reduction and de-intercalation upon oxidation as shown in Fig. 5.[37] Mo^{+6} ions in MoO$_3$ are reduced to Mo^{+5} ions upon intercalation of charges in the film. During deintercalation of charges, the Mo^{+5} sites in the film are oxidized to Mo^{+6}. The absence of the sharp oxidation and reduction peaks (Figs. 3(a) and 3(b)) during CV cycling may be attributed to the repeated reversible phase changes for Mo^{+5} to Mo^{+6} and that no separate phase is formed on the surface. It is observed from cyclic voltammetry analysis that heat-treated films of molybdenum oxide exhibit sharper oxidation peaks (Figs. 3 (c) and 3(d)) where the Mo^{+5} sites in the film are oxidized to Mo^{+6}. During the above-mentioned processes of intercalation and

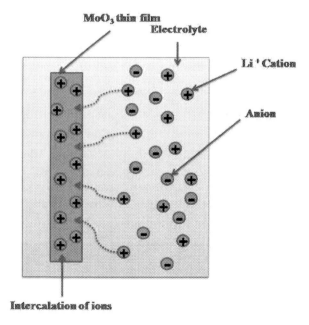

Fig. 5. Energy storage mechanism illustration of intercalation capacitance.

deintercalation of Li⁺ in MoO₃, the following electrochemical redox reaction is taking place:

$$MoO_3 + xLi^+ \leftrightarrow Li_xMoO_3. \qquad (1)$$

Li$_x$MoO$_3$ formed in the above redox reaction is the lithium-molybdenum bronze. The presence of multiple peaks in annealed films can be attributed to Li⁺ insertion at energetically distinct reaction sites within the molybdenum oxide thin films.[38]

The cyclic voltammograms reveal the electrochemical behavior of molybdenum oxide thin films and hence these thin films can find suitable applications in energy storage devices owing to their capacitive behavior. The cyclic voltammetry analysis clearly depicted the reversibility of the annealed thin films of molybdenum oxide within the given voltage window. The areal capacitance measured from the CV plots is given in Fig. 6. The slope of the plot between scan rate

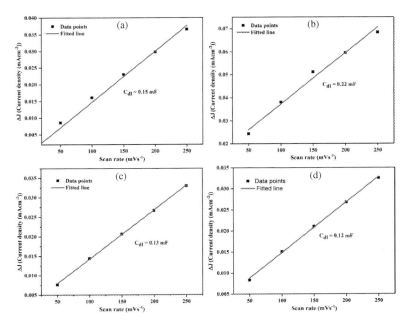

Fig. 6. Plot of peak current density versus potential scan rate for MoO₃ thin films: (a) as-deposited, (b) annealed at 150°C, (c) annealed at 250°C and (d) annealed at 350°C.

(mV/s) and the sum of anodic and cathodic currents at a fixed voltage window estimates the areal capacitance of the synthesized sample, and electrochemical active surface area (ECSA).[36] The ECSA estimated for the films is 3.75 cm^{-2} (as-deposited), 5.5 cm^{-2} (150°C annealed), 3.25 cm^{-2} (250°C annealed) and 3.0 cm^{-2} (350°C annealed) respectively. Only the sample annealed at 150°C exhibits higher areal capacitance due to a higher number of active sites (ECSA) than those annealed at relatively higher temperatures. The morphological study of thin films showed the formation of hollow tube like nano-structures at annealing temperature of 150°C which provides more active sites and hence resulted in higher areal capacitance of these films. The specific capacitance calculated for the as-deposited film is 0.179 F/g while it is 0.256 F/g for the film annealed at 150°C, 0.101 F/g for the film annealed at 250°C and 0.099 F/g for the film annealed at 350°C.

3.3.3. *Electrochemical impedance spectroscopy*

Figure 7 presents the Nyquist impedance plots in the frequency range 100 KHz to 100 MHz.

The results represent the real component of the electrical impedance of the films versus its imaginary component. The results revealed the smaller charge transfer resistance in the fabricated films. The series equivalent resistance is nearly the same in all the samples due to equal concentration of the electrolyte solutions. The plots being nonvertical and inclined to Z-axis at 45° indicates the capacitive behavior of the molybdenum oxide films. The deviation of the pattern to fit an electric circuit is attributed to the surface roughness, inhomogeneity and the presence of some constant phase element in the sample.[39] The annealing temperature has very little difference on the EIS spectra. The only difference observed is that the sample annealed at 150°C shows the lower resistance and accordingly exhibits the higher ESCA and pseudocapacitance.

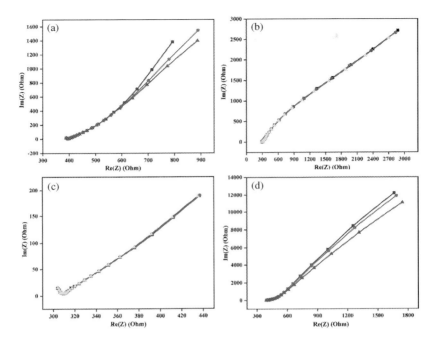

Fig. 7. Nyquist impedance plot of the MoO$_3$ thin films: (a) as-deposited, (b) annealed at 150°C, (c) annealed at 250°C and (d) annealed at 350°C.

4. Conclusions

This paper depicts the cyclic voltammetry and impedance spectroscopy studies performed on the as-deposited and annealed films of MoO$_3$. The studies clearly revealed the capacitive behavior of the MoO$_3$ thin films enabling the use of these films in various energy storage devices. The cyclic voltammetry analysis revealed that heat-treated films of molybdenum oxide exhibited sharper oxidation peaks. Upon heat treatment, the structure of the films changed from amorphous to crystalline, which being more structurally ordered exhibited higher number of active sites for activation during CV process. The CV plots also showed that the areal capacitance increased with increasing scan rates, which confirmed the enhanced charge storage capability of the synthesized films of molybdenum oxide. EIS indicated the capacitive behavior of the molybdenum oxide films. The deviation of the pattern to fit an electric circuit was attributed to the surface roughness, inhomogenity and the presence of some constant phase element in the sample.

References

1. J. Zhao and A. F. Burke, *J. Energy Chem.* **59**, 276 (2021).
2. X. Feng, X. Shi, J. Ning, D. Wang, J. Zhang, Y. Hao and Z. S. Wu, *EScience* (2021).
3. W. Tang, L. Liu, S. Tian, L. Li, Y. Yue, Y. Wu and K. Zhu, *Chem. Comm.* **47**, 10058 (2011).
4. X. Zhang, X. Zeng, M. Yang and Y. Qi, *ACS Appl. Mater. Interfaces* **6**, 1125 (2014).
5. Q. Qu, P. Zhang, B. Wang, Y. Chen, S. Tian, Y. Wu and R. Holze, *J. Phys. Chem. C* **113**, 14020 (2009).
6. C. An, Y. Zhang, H. Guo and Y. Wang, *Nanoscale Adv.* **1**, 4644 (2019).
7. W. Deng, X. Ji, Q. Chen and C. E. Banks, *RSC Adv.* **1**, 1171 (2011).
8. M. Vasilopoulou *et al.*, *J. Am. Chem. Soc.* **134**, 16178 (2012).
9. Y. Zhao, A. M. Nardes and K. Zhu, *Appl. Phys. Lett.* **104**, 213906 (2014).
10. F. Wang, X. Qiao, T. Xiong and D. Ma, *Org. Electron.* **9**, 985 (2008).
11. M. M. Alsaif, M. R. Field, B. J. Murdoch, T. Daeneke, K. Latham, A. F. Chrimes, A. S. Zoolfakar, S. P. Russo, J. Z. Ou and K. Kalantar-Zadeh, *Nanoscale* **6**, 12780 (2014).
12. A. A. Bessonov, M. N. Kirikova, D. I. Petukhov, M. Allen, T. Ryhänen and M. J. Bailey, *Nat. Mater.* **14**, 199 (2015).
13. B. Mendoza-Sa´nchez, T. Brousse, C. Ramirez-Castro, V. Nicolosi and P. S. Grant, *Electrochim. Acta* **91**, 253 (2013).
14. W. Wang, J. Qin, Z. Yin and M. Cao, *ACS Nano* **10**, 10106 (2016).
15. T. Ivanova, A. Szekeres, M. Gartner, D. Gogova and K. Gesheva, *Electrochim. Acta* **46**(13–14), 2215 (2001).
16. A. Arfaoui, S. Touihri, A. Mhamdi, A. Labidi and T. Manoubi, *Appl. Surf. Sci.* **357**, 1089 (2015).
17. M. Dhanasankar, K. Purushothaman and G. Muralidharan, *Appl. Surf. Sci.* **257**, 2074 (2011).

18. M. Rouhani, J. Hobley, G. S. Subramanian, I. Y. Phang, Y. L. Foo and S. Gorelik, *Sol. Energy Mater. Sol. Cells* **126**, 26 (2014).
19. A. L. Fernandes Cauduro, Z. E. Fabrim, M. Ahmadpour, P. F. Fichtner, S. Hassing, H. G. Rubahn and M. Madsen, *Appl. Phys. Lett.* **106**, 202101 (2015).
20. A. Quintana, A. Varea, M. Guerrero, S. Suriñach, M. Baró, J. Sort and E. Pellicer, *Electrochim. Acta* **173**, 705 (2015).
21. C. Julien, A. Khelfa, O. Hussain and G. Nazri, *J. Cryst. Growth* **156**, 235 (1995).
22. H. Martínez, J. Torres, M. Rodríguez-García and L. L. Carreño, *Phys. B: Condens. Matter.* **407**, 3199 (2012).
23. B. Han, M. Gao, Y. Wan, Y. Li, W. Song and Z. Ma, *Mater. Sci. Semicond. Process.* **75**, 166 (2018).
24. S. Touihri, A. Arfaoui, Y. Tarchouna, A. Labidi, M. Amlouk and J. Bernede, *Appl. Surf. Sci.* **394**, 414 (2017).
25. P. K. Parashar and V. K. Komarala, *Thin Solid Films* **682**, 76 (2019).
26. P. Chelvanathan, K. S. Rahman, M. I. Hossain, H. Rashid, N. Samsudin, S. N. Mustafa, B. Bais, M. Akhtaruzzaman and N. Amin, *Thin Solid Films* **621**, 240 (2017).
27. P. Singh, A. Singh and A. Thakur, *J. Mater. Sci.: Mater. Electron.* **30**, 3604 (2019).
28. P. Singh, A. Singh, J. Sharma, A. Kumar, M. Mishra, G. Gupta and A. Thakur, *Phys. Rev. Appl.* **10**, 054070 (2018).
29. S. Kumari, P. Singh, H. Singh, K. Singh, A. Kumar, S. Kumar and A. Thakur, *J. Mater. Sci.: Mater. Electron.* **32**, 24990 (2021).
30. A. Kanwade *et al.*, *J. Energy Storage* **55**, 105692 (2022).
31. R. Sivakumar, K. Shanthakumari, A. Thayumanavan, M. Jayachandran and C. Sanjeeviraja, *Surf. Eng.* **25**, 548 (2009).
32. J. Li, Q.M. Yang and I. Zhitomirsky, *J. Power Sources* **185**, 1569 (2008).
33. R. Pujari, V. Lokhande, V. Kumbhar, N. Chodankar and C. D. Lokhande, *J. Mater. Sci.: Mater. Electron.* **27**, 3312 (2016).
34. Y. Jiang and J. Liu, *Energy Environ. Mater.* **2**, 30 (2019).
35. T. Brousse, D. Bélanger and J. W. Long, *J. Electrochem. Soc.* **162**, A5185 (2015).
36. R. A. Mir, S. Upadhyay, R. A. Rather, S. J. Thorpe and O. Pandey, *Energy Adv.* **1**, 438 (2022).
37. S. Khalate, R. Kate, H. Pathan and R. Deokate, *J. Solid State Electrochem.* **21**, 2737 (2017).

38. T. M. McEvoy, K. J. Stevenson, J. T. Hupp and X. Dang, *Langmuir* **19**, 4316 (2003).
39. K. Ojha, M. Sharma, H. Kolev and A. K. Ganguli, *Catalysis Sci. Technol.* **7**, 668 (2017).

Chapter 6

V$_2$C-based lithium batteries: The influence of magnetic phase and Hubbard interaction

Jhon W. González [*,§], Sanber Vizcaya [†] and Eric Suárez Morell [‡,§]

Grupo de Simulaciones, Departamento de Física
Universidad Técnica Federico Santa María
Casilla Postal 110V, Valparaíso, Chile
[*]jhon.gonzalez@usm.cl
[†]*sanberjosevizcaya@gmail.com*
[‡]*eric.suarez@usm.cl*

MXenes are a family of two-dimensional materials that could be attractive for use as electrodes in lithium batteries due to their high specific capacity. For this purpose, it is necessary to evaluate magnitudes such as the lithium adsorption energy and the magnitude of the open-circuit voltage for different lithium concentrations. In this paper, we show through first-principles calculations that in a V$_2$C monolayer, we must consider the high correlation between the electrons belonging to vanadium to obtain correct results of these quantities. We include this correlation employing the Hubbard coupling parameter obtained by a linear response method. We found that the system is antiferromagnetic and that the quantities studied depend on the magnetic phase considered. Indirectly, experimental results could validate the theoretical value of the Hubbard parameter.

Keywords: MXenes; DFT; magnetic phase; 2D materials; batteries.

1. Introduction

The family of 2D transition metal carbides, carbonitrides, and nitrides, collectively referred to as MXenes is another emerging

---

[§]Corresponding authors.
To cite this article, please refer to its earlier version published in the Functional Materials Letters, Volume 16(8), 2340023 (2023), DOI: 10.1142/S1793604723400234.

family of 2D materials that has received growing interest due to their interesting physical and chemical properties. Among the potential applications of this family of materials are their use as the basis for the manufacture of electrodes for lithium batteries.[1,2] The feasibility of materials as ion battery electrodes is characterized by the charge capacity — related to the energy stored per unit formula — and by the open-circuit voltage (OCV) — related to the change in the chemical potential of the electrode as a function of ion concentration.[3] In particular, the V_2C is a promising 2D material for Li-ion electrodes due to its high theoretical specific capacity, estimated at 940 mAh/g. The V_2C capacity is higher than other members of the MXenes family and carbon-based electrodes.[4] Figure 1 shows a top and side view of a V_2C monolayer.

However, the magnetic character of V_2C ground state is debated. Due to the strongly localized nature of the d-electrons in transition metal atoms, the compounds involving elements like vanadium are challenging to model using *ab-initio* techniques. Electron correlation effects may lead to parameter-dependent atomic structures, electronic bandgap, and magnetic phases.[5] In some calculations, the ground state phase is considered non-magnetic (NM),[6] while in others, the authors claim it is antiferromagnetic (AF).[7] The matter is not trivial, and as we will show, the calculations with different

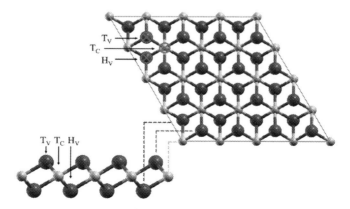

Fig. 1. Schematic view showing top and side view. The lithium adsorption sites are labeled.

magnetic phases lead to different results in the quantities associated with battery performance.[8] The electronic correlation effects on the vanadium atoms are strong, and when they are considered in *ab-initio calculations*, physical properties such as lattice constant, interatomic distances, and adsorption properties are modified. It also impacts the magnetic phase of the ground state of the system.

In this paper, we evaluate, using first-principles calculations, the impact of the U-Hubbard (DFT + U) term in the electrochemical properties of the monolayer V_2C as an electrode in a lithium-ion-based battery. First, we calculated the U-Hubbard value for the vanadium atoms in V_2C using the linear response method.[9] The obtained value depends on the density functional theory (DFT) code employed. We obtained a value of 5.0 eV for Quantum-ESPRESSO (QE) and 4.1 eV for VASP.[10,11] However, despite the difference in the values, we found the same stability order, similar magnetic moments and charges, and in particular similar ferromagnetic (FM)–AF energy differences in both codes.

Finally, using the U value obtained, we calculate the adsorption energy and the OCV in a 4×4 supercell as a function of the lithium concentration. We study the different adsorption sites of lithium atoms in V_2C and find that regardless of the magnetic character of the V_2C, the lithium tends to be adsorbed on top of the carbon atom. Our results show agreement with recent experimental results and validate the use of the U parameter to obtain the electrochemical properties of MXenes.[12] In that sense, from measurable, battery-related properties, the magnetic state of the V_2C could be inferred and used as an indirect validation of the U-Hubbard parameter determined by theoretical methods.

2. Methodology

The systems were modeled within the DFT using the plane-wave approximation implemented in VASP and QE. In both codes, we used the generalized gradient approximation (GGA) proposed by Perdew–Burke–Ernzerhof (PBE),[13] and we included the van der Waals correction using the DFT-D3 correction.[14]

We performed a thorough calculation, with and without the U-Hubbard term. We compute the U-Hubbard term using the linear response approach[9] and compare it with previous reports.[15-18] The U-Hubbard term is the computationally cheapest addition that can be used to include the correlation effects of electrons in d- and f-orbitals, and its use allows a better fit of theory to experimental measurements. However, the specific U-Hubbard value must often be determined empirically only when adequate experimental data (electronic bandgap, magnetic character, and others) are available.[19] The linear response method determines the optimal value using the differences between the derivative of the screened and bare energies with respect to the occupation N_i of the n-localized states (d-orbitals) at the atomic site i upon a perturbation α, as follows[9,20,21]:

$$U = \left(\frac{\partial N_i^{\text{SCF}}}{\partial \alpha_i}\right)^{-1} - \left(\frac{\partial N_i^{\text{NSCF}}}{\partial \alpha_i}\right)^{-1}. \quad (1)$$

In practical terms, the linear response methodology involves a small perturbation α_i to the atomic site i, acting as a Lagrange multiplier to maximize the functional to the d-orbitals belonging to one of the vanadium atoms to excite charge fluctuations in these orbitals and solve the self-consistent Kohn–Sham equations to obtain perturbed occupations.[19]

For vanadium atoms in V_2C-monolayer, in QE, we use the Löwdin-orthogonalized atomic wave functions (orthoatomic option), we find a linear response $U_{\text{QE}} = 5.0$ eV. Whereas, with VASP, we find a linear response value of $U_{\text{VASP}} = 4.1$ eV. Despite the differences between the two implementations, there is an agreement between the calculated quantities: relative energies, stability order, structural properties, adsorption properties, charges, and magnetic moments. This difference between U_{QE} and U_{VASP} is related to differences in the implementation of correlation effects in both codes. VASP uses the projectors of the pseudopotentials (PAW type mostly) to define the atomic occupations to be used in the Hubbard correction. QE atomic wave functions are instead primarily used to project Kohn–Sham states. An in-depth discussion of the differences in DFT+U implementation between VASP and QE can be found in Ref. 22.

For the calculations with QE, we use a kinetic energy cutoff for the wave functions of 80 Ry (~1090 eV) and 650 Ry (~8844 eV) for charge density. In VASP calculations, we set the cutoff energy to 520 eV. In both codes, we use a dense 21 × 21 × 1 k-grid to obtain the magnetic order. We extended the system to a 4 × 4 supercell with a 3 × 3 × 1 k-grid to perform Li-adsorption calculations.

In order to test several concentrations of lithium atoms in the V_2C monolayer, we need to work with a supercell. Based on previous works,[6,7] we extended our cell four times in the in-plane direction; we will refer to this cell as 4 × 4 cell. The order of stability referred to as the magnetic phase is the same in the primitive cell (1 × 1) as in the 4 × 4 supercell. After relaxing the 4 × 4 cell, we found a crystallographic phase with lower energy than the 1 × 1 primitive cell. This effect is especially noticeable when we consider FM solutions.

We calculated the average absorption energy (E_{ads}) of N lithium atoms in the 4 × 4 supercell using:

$$E_{ads} = \frac{E_{Li_{x_i}V_2C} - E_{V_2C} - NE_{Li}}{N}, \qquad (2)$$

where E_{V_2C} refers to the total energy of pristine V_2C layer, $E_{Li_{x_i}V_2C}$ is the energy of the system with N lithium atoms that correspond to a concentration x_i of lithium ions adsorbed on the V_2C monolayer, and E_{Li} refers to the chemical potential of a lithium-ion. The value of E_{Li} is obtained by considering one Li atom as a free-ion (gas phase). With this definition, more negative values represent higher adsorption.

Another parameter used to assess the theoretical performance of lithium-ion batteries is the OCV profile. In practical terms, the OCV is associated with the change of the chemical potential of the electrode as the lithium number changes, and it is related to the slope of the formation energy ($E_{form} = NE_{ads}$) as a function of lithium concentration. The OCV profile of ion adsorption on a surface can be calculated from the change in the Gibbs free energy of the system. When neglecting the changes in volume and entropy, the OCV expression simplifies to a difference in total energies,[23,15,24] as

$$\text{OCV} \approx \frac{E_{\text{Li}_{x_2}\text{V.2C}} - E_{\text{Li}_{x_1}\text{V.2C}} - (x_2 - x_1)E_{\text{Li}}}{(x_2 - x_1)e}, \quad (3)$$

where e is the electron charge, $E_{\text{Li}_{x_i}\text{V.2C}}$ is the total energy of the V_2C with a concentration of x_i lithium ions, and E_{Li} is the chemical potential of lithium ion.[25] Experimentally E_{Li} is a reference point used to measure the OCV in lithium batteries. Lithium foils are usually used as a reference electrode and auxiliary electrodes in experimental setups.[12] Therefore, in our calculations, we employed the E_{Li} in a BCC crystal.

Another relevant parameter to characterize any active material as an electrode for batteries is the specific capacity, also called theoretical gravimetric reversible capacity.[26] The specific capacity is the amount of charge stored in a material per mass unit. It is calculated from the chemical composition of the material storing the charge.[25] The specific capacity indicates the amount of charge stored per unit weight of the formula. The experimental value is obtained from a voltage-time curve in a galvanostatic cycle, and the theoretical specific capacity reads

$$C = \frac{NF}{M_w}, \quad (4)$$

where N is the number of ions, F is the Faraday constant, and M_w is the molecular weight of the electrode.

3. Results

3.1. *Stability order*

First, we proceed to obtain the crystallographic phase of the V_2C monolayer; due to the magnetic nature of vanadium atoms, we must consider three magnetic phases, whether the system is NM, FM, or AF. The V_2C monolayer has a crystallographic structure with a lattice similar to that of MoS$_2$ with an arrangement of 1T of the three atoms in the unit cell.[27,28] However, in MXenes[5] and some dichalcogenides,[29-31] the high electronic correlation distorts the unit cell, creating superstructures. This crystallographic phase is known as the charge density wave (CDW) phase due to the appearance in several of these structures of this type of collective excitation. Such structural distortion is stronger in FM solutions and yields finite-size effects. Later, we relax the 1T structure, and from that cell, we build a 4 × 4 supercell. The relaxation of the FM solution resulted in a crystallographic phase of 16.5 meV per formula unit lower in energy than the more symmetric 1T phase. We use this 4 × 4 super-structure to determine the magnetic phase and, subsequently, study the magnitudes associated with the performance of monolayer V$_2$C as a battery component.

We show now how the magnetic phase depends on the electronic correlation in the material; this effect is included through the parameter U of the Hubbard interaction. The magnetic moment depends on the value of U. For $U = 0$, the magnetic moment is zero, so the ground state is NM, and the solutions with magnetic moment (FM or AF) are 1 eV per formula unit above. It was necessary to force the magnetic solutions, so we have not included them in this paper.

For values of U-Hubbard different from zero, the magnetic solutions are energetically more stable than the NM solution. Despite the differences in the used value of U in VASP or QE, in both codes, the AF state is the ground state. Figure 2 shows the evolution of the stability order as a function of the U-Hubbard parameter. For small values of U-Hubbard, the magnetic moment of vanadium atoms tends to zero. For $U = 0$ eV, the NM solutions are dominant; for $U = 1.0$ eV, the magnetic moment per vanadium atom is less than one Bohr magneton 1 m_B; and for $U > 4$ eV the magnetic moment is higher than 2 m_B. The slight difference between NM/FM/AF solutions can be understood by the low magnetic moment observed for solutions with $U < 2$ eV. Although V$_2$C monolayers have been synthesized recently using different experimental techniques,[12,32] to the best of our knowledge, the AF character of the monolayer has not been tested or refuted yet.

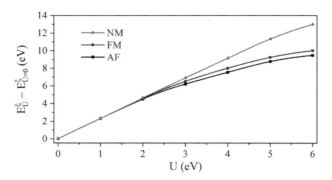

Fig. 2. Total energy as a function of the U-Hubbard term. All energies were referenced to the energy of the same NM solution ($U = 0$). ξ labels the magnetic phase (AF, FM, or NM). Black dots represent the AF solution, blue dots are for the FM solution, and red dots represent the NM solution.

As an additional comment to Fig. 2, we would like to note that when long-range van der Waals interactions are not considered, although the stability order is the same, the NM and FM solutions are closer in energy. For instance, for $U = 6$ eV, the difference in energy $E_{FM} - E_{NM}$ with van der Waals interaction is ~3 eV, and the same difference without van der Waals interaction is only 0.8 eV. We hypothesize that such an effect may be due to the small changes in structure induced by the van der Waals forces, further analysis is beyond the scope of this paper.

In Table 1, we summarize the main structural parameters for the two crystallographic structures studied, the 1T (1×1) and the 4×4 supercell for the different magnetic phases. The 4×4 lattice constant is divided by 4 for better comparison. The AF solution has the highest lattice constant (a_0), but the layer height (Δ_z) is the lowest. Finite-size effects can be seen in the structural parameters of the FM solution; the variation in height could be enough to be experimentally measured.

We found also, in agreement with previous reports,[16][18,33] that V_2C monolayer is metallic regardless of the value of the U parameter employed.

Table 1. Averaged structural parameters from the 4 × 4 and — in parenthesis — the 1T (1 × 1) cell. a_0 is the scaled lattice constant — divided by 4 in the 4 × 4 supercell — Δ_z is the layer height, $d_{V\,V}$ is the V–V first inter-plane neighbors distance, and $d_{V\,C}$ is the V–C first neighbors distance.

	a_0 (Å)	Δ_z (Å)	$d_{V\,V}$ (Å)	$d_{V\text{-}C}$ (Å)
NM	2.881	2.191	2.751	1.992
$U = 0$	(2.871)	(2.176)	(2.742)	(1.988)
AF	3.548	1.817	2.738	2.241
$U \neq 0$	(3.548)	(1.816)	(2.738)	(2.241)
FM	3.265	2.188/2.203	2.895/2.921	2.195/2.218
$U \neq 0$	(3.252)	(2.181)	(2.903)	(2.204)

3.2. Lithium adsorption

Next, we will explore the impact of the U-Hubbard parameter on the absorption energy of lithium at V_2C and whether there are changes in the variables associated with the description of the performance of an ion battery (OCV and storage capacity). On the one hand, adsorption energies are measured experimentally using chemisorption experiments, and experimental surface characterization techniques are used to identify adsorption configurations.[34,35] On the other hand, the characterization of V_2C-based lithium batteries can determine the OCV from which the binding energy for multiple lithium ions can be estimated.[36]

We identify three preferred atomic positions (details in Fig. 1): lithium on top of a carbon atom (T_C), lithium on top of a vanadium atom of the inner face (T_V), and lithium on top of a vanadium atom of the outer face (H_V). We calculated for each adsorption site, the total energy of the configuration including the Li-ion and for every possible magnetic phase: NM ($U = 0$ eV), AF ($U \neq 0$), and FM ($U \neq 0$). The preferred adsorption site, no matter the magnetic phase is the T_C (Li on top of C). In Fig. 3(a), we represent the energy difference between the different adsorption sites for each

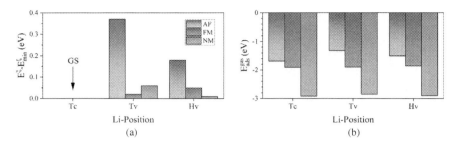

Fig. 3. Lithium adsorption energies calculated with E_{Li} in gas phase. In (a), the total energy relative to the ground state of each magnetic configuration, E_{min}^{ξ}, where ξ labels the magnetic phase. In (b), the adsorption energy is defined in Eq. (2). Black for AF ($U \neq 0$), blue for FM ($U \neq 0$), and red for NM ($U = 0$).

magnetic phase. We define E_{min}^{ξ} as the energy for each phase ξ with a Li-ion at the T_C adsorption site and then separately compare the energy of the system for the other two adsorption sites for each phase ($E^{\xi} - E_{min}^{\xi}$). The height of these bars indicates the barrier a Li-ion should face when migrating from one adsorption site to the other. Remarkably, the difference in adsorption energies between the different meta-stable sites is the largest for the AF case, indicating higher mobility barriers. For the other two phases, the difference in energy is more negligible, although the solutions have higher energy than AF.

In Fig. 3(b), we show the adsorption energy for each magnetic phase and for each site. This value indicates the feasibility of the V_2C monolayer as a battery, the smaller (more negative) the value the more difficult it is to extract the Li-ion from the system, and the optimal value should be around 1 eV. The adsorption energies are calculated by using Eq. (2). All calculations were performed in the 4 × 4 cell and Li in the gas phase was used for the energy of the Li-ion, E_{Li}.

The adsorption energy also depends on the magnetic phase of the material. The AF configuration ($U \neq 0$) has an adsorption energy of $E_{ads} = -1.5$ eV, followed by the FM solution ($U \neq 0$) with $E_{ads} = -1.7$ eV and finally the NM solution ($U = 0$) with $E_{ads} = -2.7$ eV. If the system were NM this value of adsorption energy would rule out the use of this material as an electrode providing an argument in favor of using U.

We also find that for the V$_2$C monolayer, when adding two Li-ions, the system prefers that both Li are on opposite faces and bonded to the same carbon atom.[16] When we consider lithium adsorption by both sides, it is energetically more favorable than the same side scenario.[16] The adsorption energy is 56.3 meV higher in the NM case ($U = 0$) and 89.7 meV higher in the AF case ($U \neq 0$).

3.3. *Lithium-battery capabilities*

Theoretical and experimental characterization of the V$_2$C monolayer as a potential electrode for lithium battery and the impact of simulation parameters is performed through OCV and charge capacity. The OCV is used to analyze the electronic energy changes in the electrode materials and estimate the battery's state of charge. The charge capacity relates the composition of the electrode to the amount of charge it can store; the storage capacity of Li-ion strongly depends on the surface functional groups changing with successive charge processes.[12]

To calculate the OCV, Eq. (3) is used for different concentrations of lithium, adding Li atoms to the 4 × 4 cell of V$_2$C in different configurations. Each concentration of Li corresponds to a specific capacity. To obtain the OCV values, we only consider Li adsorption at the T$_C$ positions and in the NM and AF phases corresponding to values of $U = 0$ and $U \neq 0$, respectively. We do not include the FM phase because, with nonzero U, the ground state is AF. Additionally, although there are energy differences in the magnetic order, the changes in absorption energy between the AF and FM phases are small, so the OCV values do not differ.

In the low concentration or highly dilute regimen, the absorption energy of the Li-ions depends on whether they are absorbed on the same face or different faces of the V$_2$C and coupled to the same carbon atom; the latter configuration is slightly more favorable, being ~5% lower in energy.[16]

Additionally, we chose the position of the lithium ions to maximize the initial distance between them in a few different configurations. Then for each concentration, we choose the one with the lowest energy. The difference in the adsorption energy of the system NM ($U = 0$) and AF ($U \neq 0$) decreases as the number of lithium adsorbed on the surface increases as can be seen from Fig. 3(a).

It is worth noticing that when several Li-ions are added to the system the convergence of the calculations is accelerated when Li atoms bonded to the same carbon are not vertically aligned, i.e., the Li–C–Li angle deviates from 180. Therefore, we introduced slight random variations in the x and y components of the Li atom positions ($\Delta_{\max} = \pm 0.1$ Å) away from the carbon position and then start the relaxation procedure.

The theoretical OCV values as a function of the concentration of lithium atoms in the V_2C monolayer are shown in Fig. 4. The value of the adsorption and formation energy depends on considering the lithium in a crystalline or a gaseous state, the OCV also depends on the chemical origin of the Li-ion, as we mentioned in Sec. 2, we consider E_{Li} the energy of a Li atom in a BCC crystal.

Note that the full stoichiometry corresponds to a specific capacity of up to 471 mAh/g. This theoretical specific capacity is in the range of the first cycle measurements of specific capacities of Li-ions

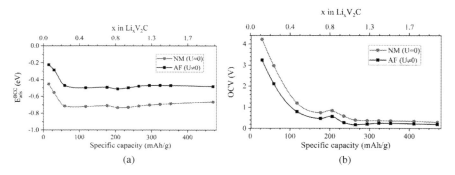

Fig. 4. Adsorption energy (a) and OCV (b) as a function of the lithium concentration. Black lines for AF ($U \neq 0$) phase and red for NM ($U = 0$). The adsorption energy and OCV values are calculated by using Eqs. (2) and (3), respectively, in both cases we consider E_{Li}, the energy of a Li atom in a BCC crystal.

batteries, with values around 500 mAh/g, decreasing in the following charge cycles.[12] Remarkably, our theoretical calculations reveal the same small bump in the OCV curve near 200 mAh/g as in the experimental measurements of Ref. 12. This point corresponds to a 14 lithium atoms concentration in the 4 × 4 supercell, which in the adsorption-energy curve (Fig. 4(a)) can be identified as a change in the trend, reaching a minimum in the curve. Such a change in the curvature appears to be associated with a change in the crystallographic structure. Any couple of lithium atoms bonded to the same carbon atom below $x = 0.7$ (Li$_x$V$_2$C) were vertically aligned; as the concentration increases, they prefer to move away from the vertical position. This shift causes some modifications in the position of Vanadium atoms, and as a result, the adsorption energy per Li atom is lowered, as we show in Fig. 4(a). Further experimental measurements are needed to confirm this unusual behavior, although the same phenomena have been already observed in the two-sided Li adsorption in MoS$_2$ monolayers, which for a critical Li-concentration, stabilize a distorted structural phase.[37]

4. Final Remarks

In this work, we studied the influence of the magnetic phase on the properties of the monolayer MXene V$_2$C as an electrode in lithium-ion-based batteries. We found that a single layer of V$_2$C behaves as an AF material and our results show that the Li-V$_2$C reaction profile strongly depends on the magnetic phase of the system and consequently on the election of the U-Hubbard parameter. Such dependence on magnetic parameters and phases leads to changes in the quantities associated with battery performance, like adsorption energies and OCV curve. In that sense, from measurable properties, the magnetic state of the V$_2$C could be inferred and even used as an indirect validation of the U-Hubbard parameter determined by the linear response method. The differences in the adsorption energy of a Li atom between the NM phase and the AF can be larger than 1 eV; the OCV curves also differ in value even though they have the same trend.

Based on experimental results, we expect that in the evaluation of MXenes and specifically of V_2C as a battery electrode, it is necessary to consider the electronic correlation to obtain adequate results of the electrochemical parameters of the material. Surprisingly, the performance of a material in a battery can give us information about the magnetic phase and the range of the U-Hubbard parameter used to model the electronic correlation effects.

Acknowledgments

ESM and JWG acknowledge financial support from ANID-FONDECYT 1221301. JWG acknowledges financial support from FONDECYT: Iniciación en Investigación 2019 Grant No. 11190934 (Chile).

ORCID

Jhon W. González ◉ https://orcid.org/0000-0003-2238-5522
Sanber Vizcaya ◉ https://orcid.org/0009-0004-9327-172X
Eric Suárez Morell ◉ https://orcid.org/0000-0001-7211-2261

References

1. R. Mishra and H. Yang, *IEEE Trans. Magn.* **57**, 1 (2021).
2. D. Sun, M. Wang, Z. Li, G. Fan, L.-Z. Fan and A. Zhou, *Electrochem. Commun.* **47**, 80 (2014).
3. Y. Chen, X. Yin, S. Lei, X. Dai, X. Xu, W. Shi, W. Liu, F. Wu and X. Cao, *Funct. Mater. Lett.* **14**, 2130011 (2021), doi: 10.1142/S1793604721300115.
4. K. Sato, M. Noguchi, A. Demachi, N. Oki and M. Endo, *Science* **264**, 556 (1994).
5. S. Bae, Y.-G. Kang, M. Khazaei, K. Ohno, Y.-H. Kim, M. J. Han, K. J. Chang and H. Raebiger, *Mater. Today Adv.* **9**, 100118 (2021).
6. M. Khazaei, M. Arai, T. Sasaki, C.-Y. Chung, N. S. Venkataramanan, M. Estili, Y. Sakka and Y. Kawazoe, *Adv. Funct. Mater.* **23**, 2185 (2013).
7. S. Zhong, B. Xu, A. Cui, S. Li, S. Liao, G. Wang, G. Liu and B. Sun, *ACS Omega* **5**, 864 (2020).

8. L. J. Bennett and G. Jones, *Phys. Chem. Chem. Phys.* **16**, 21032 (2014).
9. M. Cococcioni and S. De Gironcoli, *Phys. Rev. B* **71**, 035105 (2005).
10. G. Kresse and J. Furthmüller, *Phys. Rev. B* **54**, 11169 (1996).
11. G. Kresse and D. Joubert, *Phys. Rev. B* **59**, 1758 (1999).
12. M. Wu, Y. He, L. Wang, Q. Xia and A. Zhou, *J. Adv. Ceram.* **9**, 749 (2020).
13. J. P. Perdew, K. Burke and M. Ernzerhof, *Phys. Rev. Lett.* **77**, 3865 (1996).
14. S. Grimme, J. Antony, S. Ehrlich and H. Krieg, *J. Chem. Phys.* **132**, 154104 (2010).
15. Y.-M. Li, Y.-L. Guo and Z.-Y. Jiao, *Curr. Appl. Phys.* **20**, 310 (2020).
16. S. Nyamdelger, T. Ochirkhuyag, D. Sangaa and D. Odkhuu, *Phys. Chem. Chem. Phys.* **22**, 5807 (2020).
17. X. Ji, K. Xu, C. Chen, B. Zhang, H. Wan, Y. Ruan, L. Miao and J. Jiang, *J. Mater. Chem. A* **3**, 9909 (2015).
18. J. Hu, B. Xu, C. Ouyang, S. A. Yang and Y. Yao, *J. Phys. Chem. C* **118**, 24274 (2014).
19. Z. Xu, J. Rossmeisl and J. R. Kitchin, *J. Phys. Chem. C* **119**, 4827 (2015).
20. H. J. Kulik, M. Cococcioni, D. A. Scherlis and N. Marzari, *Phys. Rev. Lett.* **97**, 103001 (2006).
21. A. Floris, S. de Gironcoli, E. Gross and M. Cococcioni, *Phys. Rev. B* **84**, 161102 (2011).
22. Y.-C. Wang, Z.-H. Chen and H. Jiang, *J. Chem. Phys.* **144**, 144106 (2016).
23. B. Akgenç, *J. Mater. Sci.* **54**, 9543 (2019).
24. F. Zhou, M. Cococcioni, C. A. Marianetti, D. Morgan and G. Ceder, *Phys. Rev. B* **70**, 235121 (2004).
25. K. Liu, B. Zhang, X. Chen, Y. Huang, P. Zhang, D. Zhou, H. Du and B. Xiao, *J. Phys. Chem. C* **125**, 18098 (2021).
26. M. Ashton, R. G. Hennig and S. B. Sinnott, *Appl. Phys. Lett.* **108**, 023901 (2016).
27. K. Kośmider, J. W. González and J. Fernández-Rossier, *Phys. Rev. B* **88**, 245436 (2013).
28. N. Cortés, L. Rosales, P. A. Orellana, A. Ayuela and J. W. González, *Sci. Rep.* **8**, 1 (2018).
29. X. Xi, L. Zhao, Z. Wang, H. Berger, L. Forró, J. Shan and K. F. Mak, *Nat. Nanotechnol.* **10**, 765 (2015).

30. D. Cho, Y.-H. Cho, S.-W. Cheong, K.-S. Kim and H. W. Yeom, *Phys. Rev. B* **92**, 085132 (2015).
31. H. Lin, W. Huang, K. Zhao, S. Qiao, Z. Liu, J. Wu, X. Chen and S.-H. Ji, *Nano Res.* **13**, 133 (2020).
32. M. Shekhirev, C. E. Shuck, A. Sarycheva and Y. Gogotsi, *Prog. Mater. Sci.* **120**, 100757 (2021).
33. A. Champagne, L. Shi, T. Ouisse, B. Hackens and J.-C. Charlier, *Phys. Rev. B* **97**, 115439 (2018).
34. K. Bhola, J. J. Varghese, L. Dapeng, Y. Liu and S. H. Mushrif, *J. Phys. Chem. C* **121**, 21343 (2017).
35. F. Wang, K. Liu, Z. Wang, J. Zhu and S. Yin, *Funct. Mater. Lett.* **14**, 2151003 (2021), doi: 10.1142/S1793604721510036.
36. U. Yorulmaz, İ. Demiroğlu, D. Çakir, O. Gülseren and C. Sevik, *J. Phys., Energy* **2**, 032006 (2020).
37. D. Nasr Esfahani, O. Leenaerts, H. Sahin, B. Partoens and F. Peeters, *J. Phys. Chem. C* **119**, 10602 (2015).

Chapter 7

Facile graphene quantum dot-anchoring strategy synthesis of single-atom iron–nitrogen electrocatalyst with enhanced ORR performance

Huinian Zhang *, Suping Jia, Ning Li, Xiaolin Shi and Ziyuan Li

School of Energy and Power Engineering, North University of China Taiyuan 030051, P. R. China
*zhanghuinian123@163.com

Single-atom catalysts (SACs), especially atomically dispersed Fe–N_x–C based SACs, hold great promise to replace Pt-based electrocatalysts for oxygen-reduction reaction (ORR). Currently, synthesizing high-activity ORR electrocatalysts with atomically dispersed Fe–N_x site structures is still challenging due to their high surface free energy, which leads to easy migration and serious aggregation. Herein, we have designed a general graphene quantum dots (GQDs)-anchoring strategy to synthesize a single-iron-atom electrocatalyst (Fe-N-GQDs/PC) applied to ORR through calcining of N-GQDs-Fe^{3+} modified porous carbon (PC) and melamine. Experiments demonstrate the N-GQDs consist of abundant oxygenated groups, which could lead to complexing metal ions and thus facilitating the formation of SACs. Furthermore, the Fe-N-GQDs/PC electrocatalyst exhibits outstanding electrocatalytic ORR activity in 0.1 M KOH media with half-wave potentials of 0.84 versus 0.85 V for Pt/C. This strategy has opened up new feasible ideas to produce SACs for electrochemical energy devices.

Keywords: Single-atom electrocatalyst; graphene quantum dot; Fe–N active site; oxygen-reduction reaction.

---

*Corresponding author.
To cite this article, please refer to its earlier version published in the Functional Materials Letters, Volume 16(8), 2340031 (2023), DOI: 10.1142/S1793604723400313.

1. Introduction

With increasing global energy demands and environmental crisis, fuel cells provide a sustainable solution in an era of climate change and global energy consumption.[1-3] The pivotal process of fuel cells is electrochemical oxygen-reduction reaction (ORR), but the sluggish kinetics of ORR severely hinders the practical application of fuel cells.[4,5] Currently, Pt-based ORR catalysts are the most efficient and commonly used materials in fuel cells. However, they face the issue of sustainability due to their prohibitive cost and limited availability. In addition, the issues of poor stability and susceptibility of fuel (e.g. CO and methanol) also need to be resolved before widespread industry implementation.[6,7] In this context, efficient, low-cost and stable noble-metal-free catalysts, for example, nonprecious metal-based nitrogen-doped carbon electrocatalysts (M–N_x–C, e.g. M = Co, Ni, Fe, etc.) hold great promise to replace the expensive Pt-based ORR catalysts.[8-10] Among M–N_x–C materials, Fe–N_x–C represents the most promising catalysts for ORR owing to its high activity.[11-15] The most frequent approaches for synthesizing Fe–N_x–C catalysts include pyrolyzing iron–nitrogen-coordinated macrocycles or the mixtures of inorganic Fe salts, C and N precursors. However, the thermal treatment may result in Fe–N_x and uncontrollable Fe or Fe-based nanoparticles becoming embedded in the catalyst simultaneously, which complicates the identification of the real active centers of the Fe–N_x–C catalyst.[16-20] Thus, developing an efficient synthesis methodology for precisely controlling active Fe–N_x centers immobilized on the nitrogen-doped carbon supports are urgently required to boosting ORR activity.

Single-atom catalysts (SACs), fecturing high catalytic activity and maximum utilization of metal atoms, are emerging as promising electrocatalysts for ORR. They have simply active species and uniform structures, which provide a great opportunity to study the complex ORR mechanism and further improve its performance.[21-23] One possible way to boost the ORR performance is to promote the transport of ORR-relevant species as well as expose more accessible active sites. Thus, construction of SACs featured by single Fe–N_x

active sites anchored on well-defined porous carbon (PC) is deemed necessary. To date, a number of strategies have been devoted to developing carbon-supported SACs. For example, atomic layer deposition (ALD), high temperature shockwave, defect anchoring, coordinate immobilization strategy and spatial confinement have been proposed to control the formation of SACs.[24-26] It should be noted that the majority of these methods need complicated equipment that are high in cost but with low yields. For this reason, developing a controllable, lower-cost and more facile method that allows the extensive utilization of SACs in the field of fuel cells is still highly desired.

Herein, we report a graphene quantum dots (GQDs)-anchoring strategy to synthesize atomically dispersed Fe–N_x on PC substrates (denoted as Fe–N-GQDs/PC) for efficient ORR through calcining of N-GQDs-Fe^{3+} modified PC and melamine. The abundant oxygen species on N-GQDs are significantly advantageous for reinforcing the dispersion of Fe^{3+}, combined with the space-confined effect brought about by PC. The excessive N-GQDs bound to PC surface will also physically isolate the N-GQDs-Fe^{3+} complex. The combination of the protection of Fe^{3+} by N-GQDs and nitrogen doping on PC brought about by the melamine can effectively restrain the migration and aggregation of Fe atoms, as well as facilitate Fe–N_x sites' generation during the pyrolysis process. Furthermore, the obtained Fe–N-GQDs/PC catalyst exhibited enhanced ORR activity in alkaline media with an $E_{1/2}$ of 0.84 V, as well as superior long-term stability and high methanol-tolerance, which is better than that of commercial Pt/C. This synthesis method can be applied to a wide variety of non-noble metal SACs in energy conversion applications.

2. Experimental Detail

2.1. *Materials*

All reagents are of analytical grade and used as received. Citric acid monohydrate, urea, melamine, sodium alginate and iron nitrate nonahydrate were all gained from Aladdin Chemicals. 20 wt.% Pt/C was purchased from Shanghai Hesen Electric Co., Ltd. Methanol and hydrochloride acid (98%) were supplied by Shanghai Macklin Biochemical Co., Ltd. Nafion solution (5 wt.%) was bought from Alfa Aesar.

2.2. Material preparation

2.2.1. Synthesis of N-doped graphene quantum dots (N-GQDs)

The N-GQDs were synthesized according to the previous report.[27] 0.90 g (3 mmol) urea and 1.05 g (1 mmol) citric acid were added into 100 mL deionized water to form a clear and homogeneous solution under stirring. Subsequently, the solution was transferred into a sealed Teflon-lined stainless steel autoclave (250 mL). The autoclave was then heated at 160°C in an electric oven for 6 h. After that, the solution was filtered through a microporous Polytetrafluoroethylene (PTFE) membrane (0.22 μm, 13 mm). The N-GQDs were obtained from the filtered solution under rotary evaporation drying and saved for subsequent use. The obtained N-GQDs can be easily re-dispersed into water.

2.2.2. Synthesis of porous carbon (PC)

To fabricate PC, 5.0 g sodium alginate was heated to 900°C for 1 h under flowing Ar gas at 4°C min^{-1}. The as-obtained PC powder was then washed with deionized water and H_2SO_4 solution (0.2 M) to remove inorganic impurities and dried at 100°C.

2.2.3. Synthesis of Fe, N co-doped graphene quantum dots PC (Fe-N-GQDs/PC)

The obtained N-GQDs (5 mg) and $Fe(NO_3)_3 \cdot 9H_2O$ (10 mg) were dissolved in deionized water (20 mL). After sonication for 30 min, the prepared PC (33 mg) was then added into the N-GQDs-Fe^{3+} system with continuous stirring for 12 h to ensure the N-GQDs-Fe^{3+}

fully adsorb on PC. After washing with water, the slurries were harvested by drying at 60°C. The remaining solid was then ground with melamine (0.6 g) to produce a homogeneous mixture. The mixture was then calcined in Ar atmosphere at 900°C for 1 h and the heating rate was 4°C min^{-1}. After being etched in HCl (1.0 M) at 80°C for 12 h to remove the freely or loosely bonded Fe-based particles, the Fe-N-GQDs/PC was obtained. Eventually, Fe-N-GQDs/PC was dried at 100°C after washing with deionized water. For comparison, N-GQDs/PC, Fe–N-GQDs and the Fe–N-PC were synthesized by the similar approach with the Fe–N-GQDs/PC sample without the addition of Fe^{3+}, PC and N-GQDs, respectively.

2.3. Characterization

UV-Vis absorption spectra were recorded by a Shimadzu UV-2450 spectrophotometer. Transmission electron microscopy (TEM, JEM-2100F) was used to characterize the structure and morphology of the sample. Aberration-corrected TEM images and HAADF-STEM images were imaged by JEM ARM200F with double aberration correctors. The N_2 adsorption measurement was performed using ASAP 2010 system, whereas the specific surface area data were obtained on the basis of the Brunauer–Emmett–Teller (BET) method. The Bruker Vertex 70 spectrometer was used to collect the FT-IR spectra. X-ray diffraction (XRD) analyses of samples were performed on a Bruker D8 ADVANCE A25 system to verify crystal structure. Fluorescence emission spectra of samples were collected by an LS-55 fluorophotometer. XPS measurements were performed on the Thermo ESCALAB 250 spectrometer. Raman spectra were recorded on a Jobin–Yvon HR-800 Raman system.

The ORR activities were tested on a PINE electrochemical system at room temperature, which was combined with high-speed rotator equipment. All electrochemical measurements were performed at the CHI 760E electrochemical station using a three-electrode system. Pt foil and an Ag/AgCl (saturated KCl) electrode were employed as the counter and reference electrodes, respectively.

The glass carbon (GC) electrode coated with catalyst worked as the working electrode. Typically, 4 mg of Fe–N-GQDs/PC catalyst was mixed with Nafion solution, which contains isopropanol, deionized water and Nafion (v/v/v = 6/4/0.1), and then sonicated for 1 h to obtain homogeneous catalyst inks (4 mg mL^{-1}). A certain volume of catalyst inks was dropped on the surface of GC electrode toward the mass loading of 0.294 mg cm^{-2} and dried prior to use. Pt/C, Fe–N-PC, the Fe–N-GQDs and the N-GQDs/PC inks were prepared by the same method. Cyclic voltammetry (CV) tests were carried out in 0.1 mol L^{-1} KOH solution at a scan rate of 50 mV s^{-1}. Before CV tests, the solution needs purging with N_2 or O_2 for at least 1 h. The linear sweep voltammetry (LSV) of RDE and RRDE tests were performed in O_2-saturated KOH solution. The working electrode was scanned cathodically (10 mV s^{-1}) at various disk-rotation speeds from 400 rpm to 2500 rpm. The long-term stability and methanol tolerance were investigated using chronoamperometry. Note that all potentials were provided versus RHE according to the Nernst equation ($E_{RHE} = E_{Ag/AgCl}+(0.059pH+0.0197)V$). The number of transfer electrons (n) during the ORR was determined using the Koutecky-Levich equation:

$$\frac{1}{J} = \frac{1}{J_L} + \frac{1}{J_K} = \frac{1}{B\omega^{1/2}} + \frac{1}{J_K}, \quad (1)$$

$$B = 0.2nFC_0(D_0)^{2/3}v^{-\frac{1}{6}}, \quad (2)$$

where J refers to measured current density, J_L and J_K refer to diffusion-limiting and kinetic current densities (mA cm^{-2}), respectively. n represents the number of electrons transferred per O_2, ω represents the angular velocity of the GC disk (rpm), F, the Faraday constant (F = 96,485 C mol^{-1}), v is the kinetic viscosity of the 0.1 mol/L KOH solution (0.01 cm^2 s^{-1}), C_0 refers to the bulk concentration of O_2 (1.2 × 10^{-6} mol cm^{-3}), and D_0 refers to the diffusion coefficient of O_2 (1.9 × 10^{-5} cm^2 s^{-1}).

For RRDE measurements, the ring electrode potential was applied at 1.2 V (vs. RHE), yield of hydrogen peroxide (H_2O_2%)

and electrons transferred number (n) were calculated using Eqs. (3) and (4):

$$n = \frac{4I_D}{I_D + (I_R/N)}, \quad (3)$$

$$H_2O_2(\%) = 200 \times \frac{I_R/N}{I_D + I_R/N}, \quad (4)$$

where I_R is the ring current density and I_D represents the disk current density. $N = 0.37$ is the collection efficiency of the Pt ring.

3. Results and Discussion

Isolated Fe atoms were successfully embedded in PC matrix and coordinated with N atoms through the graphene QD-anchoring strategy. As shown in Fig. 1(a), biomass sodium alginate-derived

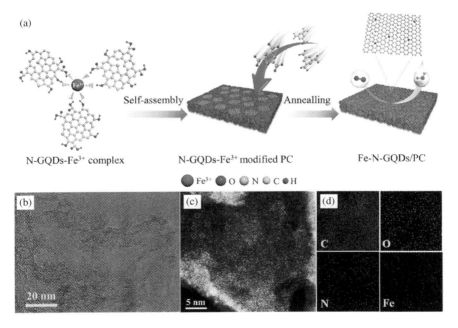

Fig. 1. (a) Schematic illustration of fabrication processes of the Fe–N-GQDs/PC electrocatalyst. (b) TEM image and (c) HAADF-STEM image of the Fe–N-GQDs/PC sample. (d) EDS maps of Fe–N-GQDs/PC.

PC was chosen as the carbon matrix to load Fe single atoms due to its earth-abundance and high specific surface area. As shown in Fig. S1(a), the PC showed a BET specific surface area of 198 m^2 g^{-1}. The pore-size distribution of PC is broad and hierarchical, which can facilitate the large exposure of more ORR active sites, as well as improve the mass transport of ions and gases [Fig. S1(b)].[28] We chose N-GQDs to tether Fe^{3+} due to its abundant –COOH and –OH functional groups on the surface (Figs. S2 and S3).[27,29] For the first step (step 1), the Fe^{3+} is complexed with N-GQDs to form a stable N-GQDs-Fe^{3+} coordination compound. The strong interaction between Fe^{3+} and O-containing groups of N-GQDs can be confirmed by the Fluorescence spectrum. Upon the addition of Fe^{3+}, the fluorescence intensity of N-GQDs decreased seriously [Fig. S4(b)]. The N-GQDs-tethered Fe^{3+} can effectively prevent the aggregation of Fe^{3+}. Therefore, achieving atomically dispersed Fe^{3+} is conducive for the synthesis of SACs. Next, the N-GQDs-Fe^{3+} coordination composite anchors on the PC surface through π-π interactions (step 2). The strong interaction between N-GQDs and PC matrix can be attributed to the π-conjugated structure of N-GQDs.[30] Finally, the N-GQDs-Fe^{3+} modified PC mixed with melamine is subjected to annealing treatment under Ar atmosphere at 900°C (step 3). During the annealing, the melamine decomposes to give N sources to form N-doped PC, which supplies enough N sites to coordinate with Fe to generate isolated Fe–N$_x$ sites anchored on N-doped PC (Fe-N-GQDs/PC).[28,31] In the absence of melamine, some crystalline Fe species can be observed on PC substrates (Fig. S5), implying that a relatively strong Fe–N bond is also conducive to overcoming the aggregation of Fe atoms. Notably, pure N-GQDs could self-crosslink in the absence of PC, which resulted in carbon nanosheet forming easily and Fe atoms aggregating into Fe-based nanoparticles (Fig. S6). This was attributed to the hydrogen bonding between oxygenated functional groups and Van der Waals bonding in N-GQDs.[30]

There is no obvious Fe-based nanoparticle in Fe–N-GQDs/PC as is observed in the high-resolution TEM image [Fig. 1(b)], which is confirmed by the XRD measurements. The XRD patterns [Fig. S7(a)] of the Fe–N-GQDs/PC sample showed a peak at 26° and a very weak peak at 43°, corresponding to the (002) and (100) diffraction peaks of graphite, respectively.[32] Except for these two peaks, no other diffraction peaks ascribed to Fe-based nanoparticles can be observed. The high magnification high-angle annular dark-filed scanning TEM (HAADF-STEM) image was provided to further confirm the existence of Fe atoms [Fig. 1(c)]. The small bright dots can be assigned to individual Fe atoms considering the heavy Fe than light C and N.[28] The elemental mapping (EDS) of Fe–N-GQDs/PC was also performed to prove the existence and homogeneous distribution of N and Fe atoms [Fig. 1(d)]. Raman spectroscopy of Fe–N-GQDs/PC reveals that the intensity of D band (1360 cm^{-1}) is higher than the G (1590 cm^{-1}) band [Fig. S7(b)], which indicates the presence of large amounts of defects in Fe–N-GQDs/PC.[33] According to a previous report, the abundant defects in Fe–N-GQDs/PC can also enhance the electrochemical performance of ORR because of the optimization of electronic configuration.[33]

The chemical composition of Fe–N-GQDs/PC was further investigated by XPS, as well as FePc for comparison to analyze the chemical bonding features of Fe and N [Fig. 2(a)]. The contents of N and Fe in the Fe–N-GQDs/PC sample determined by XPS were 5.50% and 0.30%, respectively [Fig. 2(b)]. The exact Fe content of Fe–N-GQDs/PC obtained from ICP measurement is determined to be 0.32 at.% (Table S1). The N 1s spectrum [Fig. 2(c)] exhibits five different N species, pyrrolic N (400.6 eV), graphitic N (401.4 eV), oxidized N (404.1 eV), pyridinic N (398.1 eV) and Fe–N (399.8 eV), respectively.[34] Moreover, the chemical state of Fe–N existing in the Fe–N-GQDs/PC resembles Fe–N$_4$ sites in FePc, which is a good reference sample to explore the possible Fe–N bonding. Among all types of N 1s, pyridinic N act as the sites to anchor Fe atoms.[11,31] Fe–N$_x$ and pyridinic N may contribute majorly to a good catalytic performance.[35] Furthermore, the content of these two N species is as high as 66.70% of total N content (Table S2). With such a high

118 H. Zhang et al.

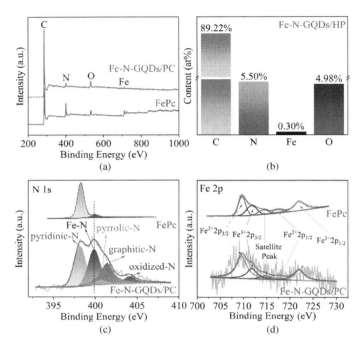

Fig. 2. (a) The XPS spectrum of Fe–N-GQDs/PC catalyst. (b) Elemental content of Fe–N-GQDs/PC catalyst obtained from XPS. (c) N 1s XPS spectrum and (d) the Fe 2p XPS spectrum of FePc and Fe-N-GQDs/PC sample.

proportion of Fe–N$_x$ and pyridinic N acting as ORR active sites, Fe–N-GQDs/PC is likely to possess excellent ORR activity. Figure 2(d) shows the high-resolution spectra of Fe 2p, which can be deconvoluted to two pairs of peaks for Fe^{2+} (709.6 and 717.7 eV) and Fe^{3+} (711.8 and 724.4 eV) with a satellite peak at 714.7 eV.[36] No clear signal of Fe^0 is detected. According to the previous report, the observation of a notable peak at ~711.8 eV in the Fe 2p spectrum [Fig. 2(d)] should correspond to Fe coordinated to N,[37,38] indicating that Fe ions are efficiently doped into the hybrid. These XPS data agree with the existence of Fe–N$_x$ coordination in Fe–N-GQDs/PC. In summary, no obvious Fe-based nanoparticles were observed in the TEM picture, and isolated Fe atoms were observed in the HAADF-STEM picture. The EDS pictures indicated the homogeneous distribution of C, Fe and N elements, and a fair proportion of Fe–N peaks existed in N 1s XPS spectrum. Based on this evidence, we confirm that abundant Fe–N$_x$ sites were successfully embedded on the Fe–N-GQDs/PC surface.

The ORR electrocatalytic behaviors of the Fe–N-GQDs/PC catalyst were tested on a three-electrode configuration in comparison with Pt/C, N-GQDs/PC, Fe–N-GQDs and the Fe–N-PC samples. The CV plots [Fig. 3(a)] of the Fe–N-GQDs/PC sample are virtually featureless in N$_2$-saturated electrolyte, while a clear cathodic peak emerged in O$_2$-saturated KOH solution, indicating good ORR activity of Fe–N-GQDs/PC. It's worth noting that the Fe–N-GQDs/PC afforded the most positive cathodic ORR peak potential, which is more positive than those of N-GQDs/PC, Fe–N-GQDs and Fe–N-PC [Fig. 3(c)], indicating Fe–N-GQDs/PC might perform superior ORR activity. The ORR activity of Fe–N-GQDs/PC was further investigated by LSV curves. It was found that Fe–N-GQDs/PC possesses excellent catalytic activity with an onset potential (E_{onset}) of 0.97 V and half-wave potential ($E_{1/2}$) of 0.84 V, which are comparable to that of commercial Pt/C (0.99 V and 0.85 V) [Fig. 3(d)]. Furthermore, the largest limited current density of Fe–N-GQDs/PC approaches that of Pt/C. The average electron transfer number (n) for the Fe–N-GQDs/PC was 3.91 according to RRDE measurements in the potential range of 0.2–0.8 V [Fig. 3(f)], consistent with the values from K–L plots [Fig. 3(b)]. In addition, peroxide yield (H$_2$O$_2$%) on Fe–N-GQDs/PC [Fig. 3(f)] during ORR was mainly below 10.0%, this further indicates Fe–N-GQDs/PC exhibit high selectivity toward a direct four-electron pathway. Moreover, the N-GQDs/PC catalyst without Fe precursor exhibited negligible ORR activity compared with the Fe–N-GQDs/PC sample [Fig. 3(d)]. However, when the Fe precursor was introduced, Fe–N-GQDs/PC showed an explosive enhancement in half-wave potential. These demonstrated that the atomically dispersed Fe–N$_x$ active sites play a leading catalyst role other than the N–C sites. The N-GQDs/PC has a higher N content than Fe–N-GQDs/PC [Table S2, Figs. S8(c) and S8(d)]. However, the ORR activity of N-GQDs/PC is much poorer than Fe–N-GQDs/PC, which further

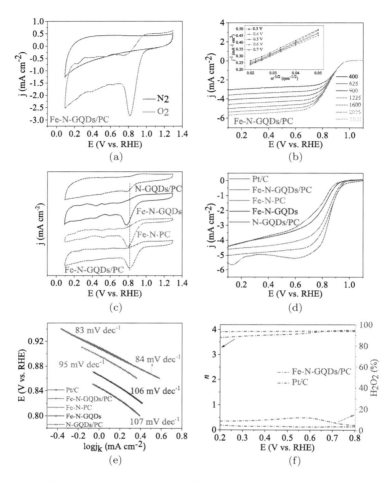

Fig. 3. (a) CV curves of Fe–N-GQDs/PC sample in N_2 (black line) or O_2 (red line) saturated 0.1 M KOH solution. (b) ORR LSVs of Fe–N-GQDs/PC sample at various rotation speeds and the inset shows K–L plots. (c) CV curves and (d) LSV plots of Fe–N-GQDs/PC, N-GQDs/PC, Fe–N-GQDs, the Fe–N-PC sample and commercial Pt/C for ORR. (e) Corresponding Tafel plots of ORR activity in d. (f) Electron transfer number (n) and the corresponding yield of H_2O_2% on Fe–N-GQDs/PC and Pt/C samples using the RRDE method.

confirms that isolated Fe–N_x active sites could efficiently improve the ORR activity. According to previous reports,[12–15] pyridine-N has the ability to bring a synergic effect to promote electrocatalytic ORR activity, while graphite-N affects the electrical conductivity of carbon matrix. Pyrrolic-N and oxidized N are generally considered as showing negligible ORR activity. By contrast, Fe–N-GQDs exhibited an $E_{1/2}$ of 0.795 V, which was not as outstanding as the performance of the Fe–N-GQDs/PC catalyst. The lower ORR activity of Fe–N-GQDs could be ascribed to the combination of the following factors. First, according to XPS N 1s spectrum analysis, the total amount of isolated active Fe–N$_x$ sites in Fe–N-GQDs is lower than that in Fe–N-GQDs/PC due to its lower total N content [Figs. S8(a,b), S11(b) and Table S2] and lower total Fe content (0.19 wt.%, Table S1). The TEM picture obviously reveals Fe aggregates coated by the carbon layers (Fig. S6), which explains the loss of Fe–N$_x$ active sites. This indicates that the absence of PC would likely lead to the loss of Fe–N$_x$ active sites. Furthermore, the BET surface area of the Fe–N-GQDs sample is only 82 m^2 g^{-1} (Fig. S9), which could further hinder Fe–N$_x$ active sites' exposure. The Tafel slope [Fig. 3(e)] of Fe–N-GQDs/PC (83 mV dec^{-1}) is smaller than that of Pt/C (84 mV dec^{-1}), which indicates its fast reaction kinetics for ORR.[39] Moreover, the chronoamperometry test (i-t) in Fig. 4(a) shows only a ~13% current drop over 9 h in Fe–N-GQDs/PC but a substantial current drop of 46% in Pt/C. After long-term durability test of Fe–N-GQDs/PC, the atomic Fe appears still well-dispersed on PC (Fig. S13). Remarkably, the current of Fe–N-GQDs/PC remains almost unchanged upon adding methanol [Fig. 4(b)], demonstrating its strong tolerance to methanol. In a

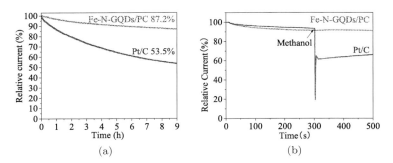

Fig. 4. (a) Durability testing of Fe–N-GQDs/PC and Pt/C at 0.7 V in O$_2$-saturated 0.1 M KOH electrolyte. (b) Methanol tolerance tests for Fe–N-GQDs/PC and commercial 20% Pt/C catalysts upon adding methanol (0.1 M).

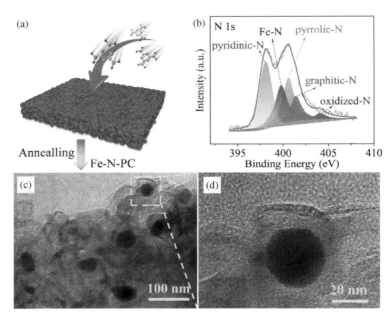

Fig. 5. (a) Schematic illustration of the synthesis process of the Fe–N-PC sample. (b) High-resolution XPS N 1s spectrum of the Fe–N-PC sample. (c) TEM picture of the Fe–N-PC and (d) metal aggregates observed in a higher magnification.

striking contrast, a substantial current drop in Pt/C is observed. The above results manifest the potential of Fe–N-GQDs/PC in fuel batteries with higher activity and better durability.

In an attempt to demonstrate the role of N-GQDs, PC was chosen as the matrix to load single-atom Fe, the other procedures were kept consistent with Fe–N-GQDs/PC except for adding N-GQDs [Fig. 5(a)], and the corresponding product was named as Fe–N-PC. Remarkably, Fe–N-GQDs/PC outperformed Fe–N-PC by 10 mV in $E_{1/2}$ [Fig. 3(d)]. This could be mainly attributed to the difference of Fe–N$_x$ active sites' amounts. The content of Fe–N$_x$ activity sites in the Fe–N-PC sample surface determined by XPS spectrum is 22.08 at.% [Fig. 5(b) and Table S2], which is much lower than that in the Fe–N-GQDs/PC sample (32.40 at.%, Table S2). The fewer Fe–N$_x$ activity sites could be ascribed to the absence of N-GQDs-tethering effect, resulting in the aggregation of Fe atoms during the calcination processes. As shown in Figs. 5(c) and 5(d), TEM pictures obviously reveal Fe-based nanoparticles' formation on PC and incorporated into the carbon layers. The formation of Fe-based nanoparticles might originate from the insufficient densities of surface O-containing functional groups on PC. The surficial O concentration in PC determined by XPS spectrum is 6.18 at.% [Fig. S10(b)], much lower than that of N-GQDs [40.11 at.%, see Fig. S3(b)]. The amount of O-containing functional groups on PC is not sufficient as anchoring sites to isolate Fe.[30] Hence, less Fe was anchored on PC surface [Fig. S11(a) and Table S1], which finally results in the limited formation of Fe–N_x active sites. When mixing N-GQDs with PC, N-GQDs can anchor on the PC surface through π–π interactions. The XPS spectrum [Fig. S10(a)] confirms that N-GQDs are covered on PC. Therefore, N-GQDs can assist PC to capture Fe atoms inhibiting the aggregation into nanocrystals [Fig. 1(c)]. Thus, the introduction of N-GQDs is essential for achieving an efficient SAC to improve ORR activity.

4. Conclusion

In conclusion, a graphene QD-anchoring strategy has been developed to synthesize atomically dispersed Fe–N_x sites anchored on the N-doped porous carbon through calcining of N-GQDs-Fe^{3+} modified PC and melamine. As a result, the synthesized Fe–N-GQDs/PC showed enhanced electrocatalytic performance for the efficient ORR with high half-wave potentials of 0.84 versus 0.85 V for Pt/C in 0.1 M KOH mediums, along with long-term stability and strong tolerance to methanol. This further broadens the synthesis method of many other SACs for energy-conversion applications.

Acknowledgments

This work is financially supported by the Fundamental Research Program of Shanxi Province (20210302123029), the National Natural

Science Foundation of China (22105181) and the Key Laboratory of Hubei Province for Coal Conversion and New Carbon Materials (Wuhan University of Science and Technology) (WKDM202202).

ORCID

Huinian Zhang ◉ https://orcid.org/0000-0003-3745-5466
Ning Li ◉ https://orcid.org/0000-0002-4046-4825

Appendix

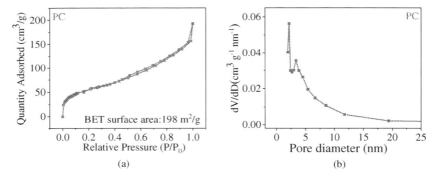

Fig. S1. (a) N_2 adsorption-desorption isotherm, (b) the corresponding pore size distribution of biomass sodium alginate derived porous carbon (PC).

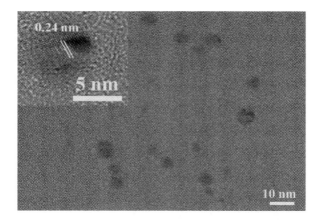

Fig. S2. HR-TEM image of the N-GQDs sample.

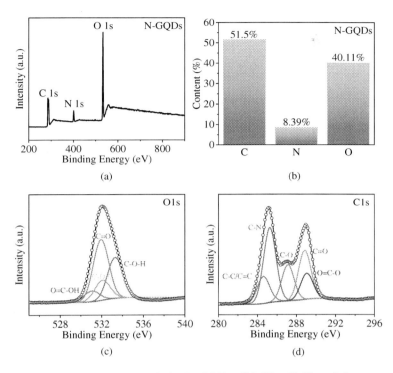

Fig. S3. (a) Survey scan of XPS for N-GQDs. (b) The C, N and O contents of N-GQDs illustrated by XPS spectrum analysis. (c) The high-resolution O 1s spectra of N-GQDs. (d) The high-resolution C 1s spectra of N-GQDs. The XPS verifies the introduction of O-functional groups in N-GQDs. The total O content in N-GQDs is around 40.11 at.%.

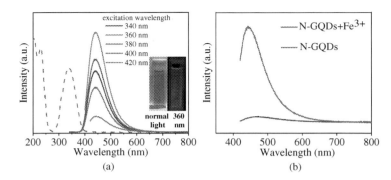

Fig. S4. (a) The UV-Vis spectrum (dash curve) and PL emission spectra (solid curves) under different excitation wavelength of N-GQDs. Insets show the optical images of the N-GQDs aqueous solution under the normal light and excited by 360 nm and (b) The Fluorescence spectrum of N-GQDs before and after addition of Fe^{3+}, respectively.

The N-GQDs solution shows yellow (Fig. S4a, inset) and two clear absorption bands at 232 nm and 340 nm (Fig. S4a, dash curve). The absorption peak at 232 nm and 340 nm are related to $\pi \rightarrow \pi^*$ transition of C=C bond and $n \rightarrow \pi^*$ transition of C=O bond, respectively. The N-GQDs solution shows a blue light (440 nm) when it excited under a 360 nm UV beam (Fig. S4a, insets). The N-GQDs shows nearly excitation-independent (Fig. S4a, solid curves).[27]

Facile graphene quantum dot-anchoring strategy synthesis 127

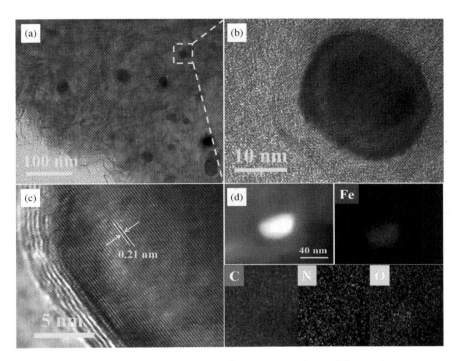

Fig. S5. (a) TEM picture of N-GQDs/PC supported Fe, (b) metal aggregates encapsulated by graphitic carbon layers observed in a higher magnification, which was synthesized by the same method as that for Fe-N-GQDs/PC, except for adding melamine, (c) high-resolution TEM image of metal aggregates in Fig. S5b and (d) the EDS-mapping images of N-GQDs/PC supported Fe (except for adding melamine).

The continuous plane lattice spacing of 0.21 nm in Fig. S5c is well consistent with the d-spacing of (220) crystallographic planes of Fe_3C or (110) planes of cubic Fe.[16,40] The EDS mapping investigation revealed that N-GQDs/PC supported Fe sample (without adding melamine) are composed of Fe, C, N, and O elements (Fig. S5d). While the Fe element intensively spread over the Fe-based nanoparticles.

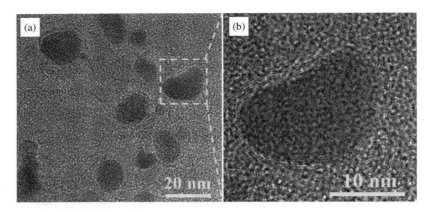

Fig. S6. (a) TEM picture of N-GQDs supported Fe and (b) metal aggregates observed in a higher magnification, which was synthesized by the same method as that for Fe-N-GQDs/PC, except for adding PC.

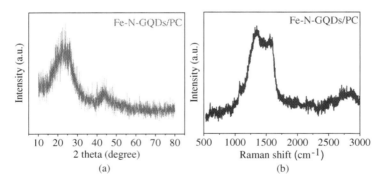

Fig. S7. (a) XRD spectra and (b) Raman spectra of the Fe-N-GQDs/PC sample.

Table S1. The Fe content (wt%) of Fe-N-GQDs/PC, Fe-N-GQDs and Fe-N-PC measured by ICP method.

	Fe-N-GQDs/PC	Fe-N-GQDs	Fe-N-PC
Fe content[wt%]	0.32	0.19	0.28

Table S2. Atomic contents of pyridinic N, oxidized N, Fe-N$_x$, graphitic N and pyrrolic N based on fine scan of N 1s of XPS and O content of Fe-N-GQDs/PC, Fe-N-GQDs, Fe-N-PC and N-GQDs/PC based on XPS analysis.

	Fe-N-GQDs/PC	Fe-N-GQDs	Fe-N-PC	N-GQDs/PC
Pyridinic N	34.30	40.74	32.18	91.96
Fe-N	32.40	26.12	22.08	—
Pyrrolic N	5.67	24.11	29.87	—
Graphitic N	27.63	8.98	15.87	8.04
Oxidized N	8.78	2.97	3.72	—
N total atomic ratio [%]	5.50	5.08	5.15	6.67
O total atomic ratio [%]	4.98	10.03	7.10	12.61

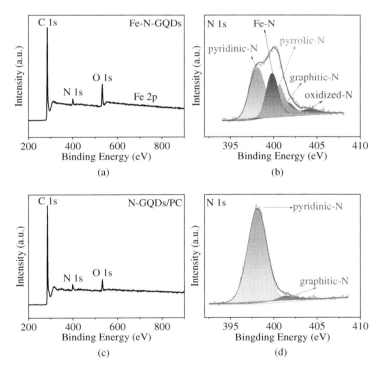

Fig. S8. (a) XPS full spectra of Fe-N-GQDs, (b) N 1s XPS spectra with deconvolution, (c) XPS full spectra of N-GQDs/PC and (d) N 1s XPS spectra with deconvolution.

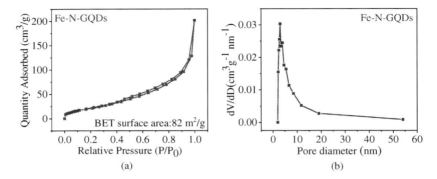

Fig. S9. (a) N_2 adsorption-desorption isotherm and (b) pore size distribution of Fe-N-GQDs.

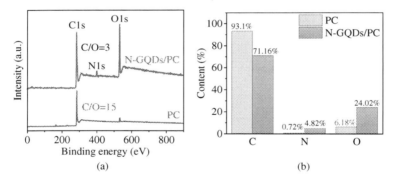

Fig. S10. (a) XPS spectrum of PC before and after addition of N-GQDs, respectively. XPS confirm that N-GQDs is covered on PC in view of the increased C/O ratio and decreased N content. (b) The C, N and O contents of PC and N-GQDs/PC illustrated by XPS spectrum analysis.

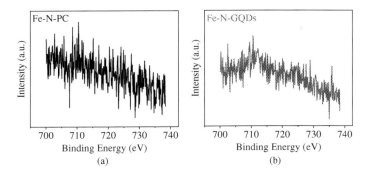

Fig. S11. (a) Fe 2p XPS spectra for the Fe-N-PC and (b) Fe 2p XPS spectra for Fe-N-GQDs.

The Fe-N-PC and Fe-N-GQDs show relatively weak peak intensity of Fe 2p XPS spectra (Fig. S11), owing to their limited Fe content.

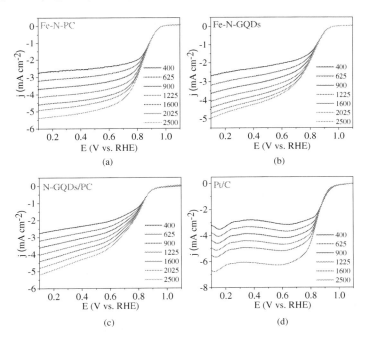

Fig. S12. LSV curves of various catalysts with various rotation rates (a) Fe-N-PC, (b) Fe-N-GQDs, (c) N-GQDs/PC and (d) Pt/C.

LSV plots of Fe-N-PC, Fe-N-GQDs, N-GQDs/PC and commercial Pt/C (20 wt% Pt) were recorded in Fig. S12. It was found that with the increase of rotation rate, the diffusion-limited current density increased. This was attributed to the short diffusion distance in KOH solution at high rotating speed.[41]

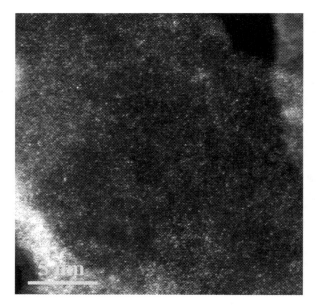

Fig. S13. The STEM image of Fe-N-GQDs/PC after long-term durability test.

References

1. G. M. Whitesides *et al.*, *Science* **315**, 796 (2007).
2. M. K. Debe *et al.*, *Nature* **486**, 43 (2012).
3. M. S. Dresselhaus *et al.*, *Nature* **414**, 332 (2001).
4. F. Y. Cheng *et al.*, *Chem. Soc. Rev.* **41**, 2172 (2012).
5. M. Lefevre *et al.*, *Science* **324**, 71 (2009).
6. L. Z. Bu *et al.*, *Science* **354**, 1410 (2016).
7. B. Lim *et al.*, *Science* **324**, 1302 (2009).
8. G. Wu *et al.*, *Science* **332**, 443 (2011).
9. J. X. Han *et al.*, *Appl. Catal. B Environ.* **280**, 119411 (2021).
10. S. Y. Chen *et al.*, *J. Am. Chem. Soc.* **144**, 14505 (2022).
11. H. N. Zhang *et al.*, *Mater. Chem. Front.* **5**, 8127 (2021).
12. Z. P. Zhang *et al.*, *Angew. Chem. Int. Ed.* **57**, 1 (2018).
13. U. Martinez *et al.*, *Adv. Mater.* **31**, 1806545 (2019).
14. C. X. Hu *et al.*, *Nano Energy* **82**, 105714 (2021).
15. G. G. Yang *et al.*, *Nat. Commun.* **12**, 1734 (2021).
16. W. J. Jiang *et al.*, *J. Am. Chem. Soc.* **138**, 3570 (2016).

17. G. J. Li *et al.*, *Energy Stor. Mater.* **56**, 394 (2023).
18. W. J. Zhai *et al.*, *Small* **2107225**, 1 (2022).
19. X. Ao *et al.*, *ACS Nano* **13**, 11853 (2019).
20. S. N. Zhao *et al.*, *Adv. Mater.* **34**, 2107291 (2022).
21. X. Luo *et al.*, *Nano-Micro Lett.* **12**, 163 (2020).
22. C. Z. Zhu *et al.*, *Adv. Energy Mater.* **8**, 1801956 (2018).
23. R. Jiang *et al.*, *J. Am. Chem. Soc.* **140**, 11594 (2018).
24. M. L. Xiao *et al.*, *Angew. Chem. Int. Ed.* **58**, 9640 (2019).
25. X. X. Wang *et al.*, *Adv. Mater.* **30**, 1706758 (2018).
26. H. Yan *et al.*, *Nat. Commun.* **8**, 1070 (2017).
27. D. Qu *et al.*, *Nanoscale* **5**, 12272 (2013).
28. H. N. Zhang *et al.*, *Angew. Chem. Int. Ed.* **58**, 14871 (2019).
29. D. Qu *et al.*, *Sci. Rep.* **4**, 5294 (2014).
30. S. Jin *et al.*, *Angew. Chem. Int. Ed.* **59**, 21885 (2020).
31. H. N. Zhang *et al.*, *Green Chem.* **20**, 3521 (2018).
32. L. Jiao *et al.*, *Angew. Chem. Int. Ed.* **57**, 8525 (2018).
33. J. J. Huo *et al.*, *J. Mater. Chem. A* **8**, 16271 (2020).
34. L. Zhao *et al.*, *Nat. Commun.* **10**, 1278 (2019).
35. R. Ding *et al.*, *Nano Res.* **13**, 1519 (2020).
36. H. L. Jiang *et al.*, *ACS Appl. Mater. Interfaces* **7**, 21511 (2015).
37. H. R. Byon *et al.*, *Chem. Mater.* **23**, 3421 (2011).
38. L. Lin *et al.*, *J. Am. Chem. Soc.* **136**, 11027 (2014).
39. Z. H. Wang *et al.*, *Adv. Funct. Mater.* **28**, 1802596 (2018).
40. S. C. Han *et al.*, *Adv. Energy Mater.* **8**, 1800955 (2018).
41. P. F. Tian *et al.*, *J. Power Sources.* **448**, 227443 (2020).

Chapter 8

Effect size of carbon micro-nanoparticles on cyclic stability and thermal performance of Na$_2$SO$_4$·10H$_2$O–Na$_2$HPO$_4$·12H$_2$O phase change materials

Zengbao Sun [*], Xin Liu [†], Shengnian Tie [*,§] and Changan Wang [‡]

[*]*New Energy Photovoltaic Industry Research Center*
Qinghai University, Xining 810016, P. R. China
[†]*School of Chemical Engineering*
Qinghai University, Xining 810016, P. R. China
[‡]*State Key Lab of New Ceramics and Fine Processing*
School of Materials Science and Engineering
Tsinghua University, Beijing 100084, P. R. China
[§]*tieshengnian@163.com*

In this paper, carbon particles with micro- and nano-particle size were synthesized through a hydrothermal reaction of glucose, namely C-1(123.1 nm), C-2(229.2 nm), C-3(335.1 nm), C-4(456.2 nm) and C-5(534.0 nm) with distinct sizes. We utilized five size carbon particles as individual fillers into the EHS matrix materials to prepare composite eutectic phase change materials (C/EHS PCMs) by melt blending technique. The impact of carbon particle size on the dispersion stability and thermal properties of Na$_2$SO$_4$·10H$_2$O–Na$_2$HPO$_4$·12H$_2$O (EHS) phase change materials was investigated. Scanning electron microscopy (SEM) and dynamic light scattering (DLS) analysis were done to analyze the diameters of carbon particles. The cryogenic-scanning electron microscopy (Cryo-SEM) analysis indicated that the carbon particles resulted in modification in the morphology of the EHS. The results of *in situ* X-ray diffraction (XRD) and Fourier-transformed infrared (FTIR) analysis showed only simple physical

---

[§]Corresponding author.
To cite this article, please refer to its earlier version published in the Functional Materials Letters, Volume 16(8), 2340032 (2023), DOI: 10.1142/S1793604723400325.

mixing between carbon particles and EHS. It is shown that adding 0.2 wt.% C-2 can decrease the supercooling degree of EHS to 1.5°C. The cyclic stability of C/EHS varies significantly depending on the size of carbon particles. The thermal conductivity of EHS increased by 42.1%, 39.9%, 14.4%, 19.5%, and 18.8% with the addition of C-1, C-2, C-3, C-4, and C-5, respectively, at a mass fraction of 0.2%. The results of differential scanning calorimetry reveal that the incorporation of C-1, C-2, C-3, and C-4 into EHS leads to an enhancement of latent heat. The latent heat capacity of EHS with 0.2 wt.% C-2 is 243.4 J·g^{-1}, and after undergoing 500 cycles of solid-liquid phase transition, the latent heat remained above 200 J·g^{-1}. Through the comprehensive analysis, the C-2/EHS composite phase change material holds significant potential for advancing building insulation and solar energy storage technologies.

Keywords: Particle size; carbon; eutectic salt; phase change material; cyclic stability.

1. Introduction

Energy is the foundation of human survival and development. With the continuous increase in global energy consumption, the vigorous development of renewable energy is necessary.[1-3] Excellent energy storage systems are of critical importance in addressing the challenges of large decentralization, unstable output, and uneven temporal and spatial distribution of renewable energy.[4] Thermal energy storage technologies include sensible heat storage, latent heat storage, and thermochemical heat storage, which are among the key technologies supporting the large-scale development of renewable energy.[5,6] Phase change materials (PCMs) attract much attention of researchers due to their advantages of high energy storage density, absence of chemical reactions, and maintain storage and release temperatures near the phase transition temperature.[7-9] Inorganic hydrate salt PCMs have the characteristics of affordability, easy availability, high energy storage density, environmental friendliness, and reusability,[10,11] which make them a kind of promising energy storage material with broad application prospects. However, the application of inorganic hydrate salts is limited by phase change temperature. Currently, the control of phase change temperature is achieved by preparation process. Yushi[12] designed eutectic hydrated salt (EHS) PCMs with different mass fractions of $Na_2CO_3 \cdot 10H_2O$–$Na_2HPO_4 \cdot 12H_2O$ and found that the phase change temperature of eutectic salt varies depending on the different fraction of hydrated salt. Dongxian[13] successfully prepared a binary eutectic salt of $Na_2SO_4 \cdot 10H_2O$–$CH_3COONa \cdot 3H_2O$ with molar ratios of 0.71:0.29 using melt blending technology, resulting in phase change temperature reduction to 28.5°C. Furthermore, almost hydrate salts exhibit problems of low thermal conductivity and serve supercooling. One of the effective approaches to alleviate the problem of supercooling is adding nucleating agents. Ning[14] successfully reduced the supercooling of $Na_2HPO_4 \cdot 12H_2O$ from 18.4°C to 1.0°C by incorporating 4 wt.% α-Al_2O_3 and 1.5 wt.% sodium alginate. Unfortunately, there is no universal method for selecting nucleating agents. Researchers just have to keep trying. Nanoparticles enhanced phase change materials[15,16] refer to novel composited PCMs produced by dispersing highly thermal conductivity nanoparticles into PCM. Many researchers[17-21] consider it to be one of the most effective techniques for enhancing the thermal conductivity of phase change materials. Xin et al.[22] added 3 wt.% mass fraction of Al/C hybrid nanoparticles to $Na_2SO_4 \cdot 10H_2O$. so that the thermal conductivity increased by 26.41%. In addition, the large specific surface area of nanoparticles can provide numerous nucleation sites for crystallization of PCM, thereby beneficial for further reducing the supercooling of PCM. Fei et al.[23] added TiO_2 nanoparticles to $CaCl_2 \cdot 6H_2O$–$MgC_{12} \cdot 6H_2O$ eutectic phase change materials. At a mass fraction of 0.3 wt.%, the thermal conductivity increased by 22.9% at liquid. The composite addition of $SrCl_2 \cdot 6H_2O$ and TiO_2 reduced the supercooling of eutectic hydrate salt by 0.3°C.

In recent years, many scholars have discovered that the addition of nanocarbon materials with high thermal conductivity (3000–6000 $W \cdot m^{-1} \cdot k^{-1}$) contributes to the thermal performance enhancement of PCM.[24,25] Meizhi et al.,[26] respectively, added 3.0 wt.% graphene nanoplatelets, multi-walled carbon nanotubes (MWCNTs), and nano-graphene to myristic acid, resulting in the thermal conductivity of PCM improved by 176.26%, 47.30%, and 44.01%, respectively. Significantly, the addition of nanoparticles not only improves the thermal conductivity of PCMs but also affects other properties such as supercooling, latent heat capacity, and phase transition temperature. Kumar[27] found that the addition of pristine MWCNTs and functionalized MWCNTs not only enhances the thermal conductivity of hydrate salt PCMs but also improves the latent heat capacity. The latent heat capacity of PCMs increased by 14.66% and 31.17% after adding 0.1 wt.% pristine MWCNTs and 0.3 wt.% functionalized MWCNTs to PCMs, respectively. The enhancement of thermal conductivity in nanoparticle-enhanced PCMs is significantly influenced by the particle material, particle size, shape, and degree of aggregation.[28,29] Gupta et al.[30] investigated the effect of carbon spheres, MWCNTs, mesoporous carbon, graphene nanoplatelets, and nano-graphite when independently added to magnesium nitrate hexahydrate (MNH) on its thermophysical properties. Results show that the carbon sphere exhibited the most significant enhancement in thermal conductivity for MNH. When adding a mass fraction of 0.5 wt.% carbon sphere, the thermal conductivity of MNH improved by 100%. In particular, due to the unique solid–liquid phase transition nature of hydrate salt phase change materials, it has a higher requirement for the dispersion stability of nanoparticles within the matrix materials. Kannan et al.[31] observed the latent heat of $LiNO_3 \cdot 3H_2O$ reduced by 8–52% after solid–liquid 1000 cycles. However, in the previous research, no work has been reported on the impact of particle size on the performance of eutectic salt phase change materials. Additionally, there is a lack of research reports on the cyclic stability of PCMs with added nanoparticles.

To remedy this knowledge gap, herein, we selected a mass ratio of 2:8 of $Na_2SO_4 \cdot 10H_2O$–$Na_2HPO_4 \cdot 12H_2O$ and a series amount of self-synthesized carbon spheres. The composited eutectic PCMs containing different sizes of nanocarbon spheres were prepared. The influences of carbon particles size on the dispersion stability and cycle life of the eutectic hydrate salt were investigated. Meanwhile, the effects of carbon particle size on the supercooling degree, thermal conductivity, and latent heat capacity in the eutectic salt PCM were furtherly investigated. This paper has positive significance for the selection of particle size into hydrate salt PCMs.

2. Experimental Procedure

2.1. Raw materials

Sodium sulfate decahydrate ($Na_2SO_4 \cdot 10H_2O$) and disodium hydrogen phosphate dodecahydrate ($Na_2HPO_4 \cdot 12H_2O$) were purchased by Shanghai Aladdin chemical Co., purity>99%, glucose was provided by Shanghai Aladdin chemical Co., purity>99%.

2.2. Sample preparation

2.2.1. Preparation of carbon particles

The synthesis method of different sizes of carbon particles could be adjusted by changing the precursor concentration of glucose solution, reaction temperature and time, as shown in Table 1. The synthesis process of C-1 carbon particles is presented below. First, 10 g of glucose was weighed and dissolved in 40 ml of water, and then the resulting solution was transferred to a Teflon-lined reactor with a capacity of 50 ml. The hydrothermal reaction was conducted at a temperature of 180°C for 9 h. Following the completion of the reaction, the product underwent multiple centrifugation cycles using water and ethanol alternatively. Subsequently, it was dried in an oven at 70°C for 48 h to obtain C-1 carbon particles.

Table 1. Synthesis parameters of different size carbon particles prepared by hydrothermal reaction.

Sample	Glucose (g)	Water (ml)	Temperature (°C)	Time (h)	Size (nm) ($n=3$)
C-1	10	40	180	9	123.1 ± 7.4
C-2	10	40	200	3	229.2 ± 6.2
C-3	1	40	200	9	335.1 ± 8.0
C-4	1	40	200	3	456.6 ± 11.7
C-5	5	40	200	6	534 ± 17.2

140 Z. Sun et al.

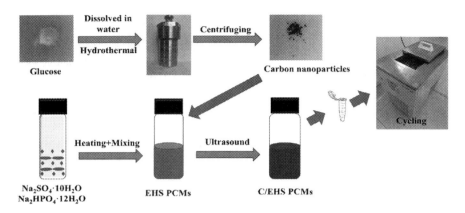

Fig. 1. Flow schematic diagram of the preparation of C/EHS PCMs.

2.2.2. Preparation of C/EHS PCMs

$Na_2SO_4 \cdot 10H_2O$ (20 wt.%) and $Na_2HPO_4 \cdot 12H_2O$ (80 wt.%) were accurately weighed and thoroughly mixed at a temperature of 50°C until the mixture in the tube achieved clarity and no solid could be found. The flow chart depicting the procedure is illustrated in Fig. 1.

2.3. Characterization

The microstructure and particle size of carbon particles were carried out using Scanning Electron Microscope (SEM), by model ZEISS Sigma 300. The microstructure of C/EHS PCMs was observed by a Hitachi Regulus8220 field electron cryogenic-scanning electron microscopy (Cryo-SEM). The surface morphology of the carbon particles was observed by a JEM-2100F Transmission Electron Microscope (TEM). The particle size distribution of carbon particles was determined using a Malvern Zetasizer Nano ZS90 nanoparticle size analyzer, employing Dynamic Light Scattering (DLS) technique. The composition of the carbon particles was analyzed using a D-max2500PC X-ray diffractometer (XRD). The test conditions were set at a range of 10–80° and a scanning rate of 8°/min. The compositions of EHS and C-2/EHS were analyzed using an in-situ X-ray diffraction technique with the Rigaku Smartlab model. The test conditions for the analysis were set at a range of 5–90° and a scanning rate of 8°/min. The chemical interactions between the carbon particles and pristine EHS in the C/EHS composites PCMs were characterized using a Bruker INVENIO-S Fourier-Transformed Infrared spectrophotometer (FTIR) in the wavenumber range of 4000–400 cm^{-1}. Differential Scanning Calorimetry (DSC) by model STA449F3–DSC200F3 was used to determine the phase transition temperatures and enthalpies of C/EHS composited PCMs. This measurement was carried out in a cyclic temperature range of −10°C to 60°C with a scan heating rate of 5°C/min under nitrogen atmosphere. The thermal stability of EHS samples was analyzed by Thermal Gravimetry Analysis (TGA), by model STA449 F3–STA449 F3 instrument in the temperature range of room temperature to 200°C with a scan heating rate of 10°C /min under nitrogen atmosphere. The thermal conductivity of the samples was determined using a TPS 2200 Hot Disk thermal constant analyzer in the liquid phase at a temperature of 45°C. The analysis was conducted using a Kapton 5465 probe under the following test conditions: 80 MW power and 40 s test time. The real-time temperature of the sample cooling process was measured using a Seitron RC-4 temperature logger. The cycling stability test of the C/EHS samples was conducted using the TMS8035-R30 high and low-temperature thermostat tank from Zhejiang Tomos Company. The cycling conditions for the test were as follows: the upper limit temperature was set at 45°C with a temperature rise duration of 25 min, and the lower limit temperature was set at 15°C with a temperature drop duration of 35 min.

3. Results and Discussion

3.1. *Size of carbon particles*

Figures 2(a)–2(c) present the SEM images of carbon particles with different particle sizes. From Figs. 2(a) to 2(e), we can see the carbon diameters of sample C1 to C5 were measured as 120 nm, 220 nm, 330 nm, 430 nm and 530 nm, respectively. The results of SEM are

142   Z. Sun et al.

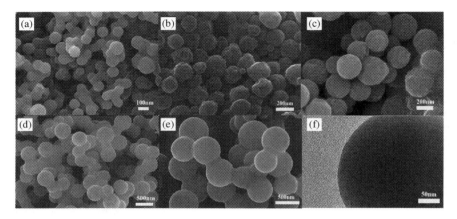

Fig. 2. SEM image of different sized carbon particles (a) C-1, (b) C-2, (c) C-3, (d) C-4, (e) C-5, (f) TEM image of C-2.

generally consistent with the particle size distribution measured using DLS, as depicted in Fig. 3. Furthermore, Fig. 2 reveals that all carbon particles exhibit a uniform and spherical shape. Samples C-2 and C-3 display individual carbon spheres without any visible agglomeration. C-1, C-4 and C-5 exhibit varying degrees of agglomeration, which may impact their dispersion within EHS PCMs. The clarity microstructure of the C-2 carbon sphere is presented in Fig. 2(f), it can be seen that the surface of the carbon sphere appears uniformly smooth, without visible holes and gaps.

3.2. Structural analysis

Figure 4(a) shows the X-ray diffraction patterns of carbon particles. It is illustrated that the nanoparticles in sample C-1 to sample C-5 show predominant peaks at 2 theta values of 23°. No other peaks indicating the presence of impurities are observed. This further confirms that the nanoparticles synthesized using the hydrothermal method with glucose are indeed carbon particles. The X-ray diffraction patterns of pristine EHS and EHS with 0.2 wt.% C-2 are shown in Fig. 4(b). The spectrum of pristine EHS contains a series of characteristic peaks both $Na_2SO_4 \cdot 10H_2O$ and $Na_2HPO_4 \cdot 12H_2O$. It is

Effect size of carbon micro-nanoparticles on cyclic stability   143

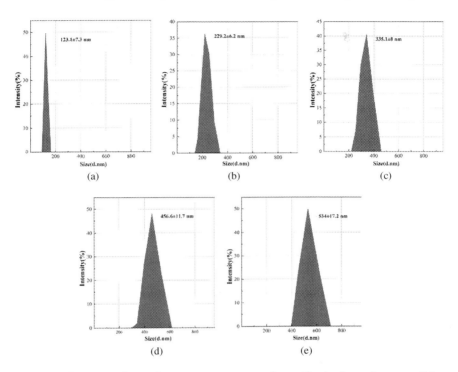

Fig. 3. Corresponding size measurements of synthesized carbon particles (a) C-1, (b) C-2, (c) C-3, (d) C-4, (e) C-5.

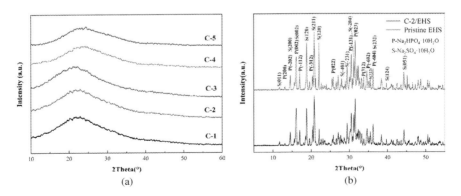

Fig. 4. XRD patterns of (a) carbon particles and (b) Pristine EHS and EHS with C-2 nanoparticles.

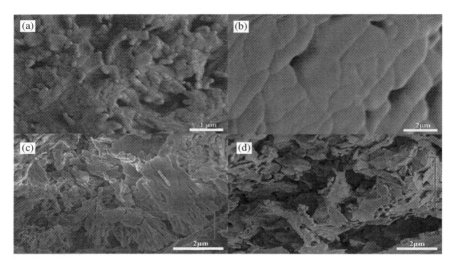

Fig. 5. SEM images of (a) Na$_2$SO$_4$·10H$_2$O, (b) Na$_2$HPO$_4$·12H$_2$O, (c) Na$_2$SO$_4$·10H$_2$O–Na$_2$HPO$_4$·12H$_2$O, (d) 0.2wt% C-2/EHS.

proved that EHS is merely a physical mixture of the two components and no chemical reaction between them. No new peaks appear in the XRD pattern of the C-2/EHS composite, confirming that the crystalline phase of pristine EHS is not affected by the addition of C-2 carbon particles, and there is no physical interaction between the dispersed carbon particles and EHS in composite PCMs.

SEM images of the pure Na$_2$SO$_4$·10H$_2$O and Na$_2$HPO$_4$·12H$_2$O in solid state were are displayed in Fig. 5(a) and 5(b). The crystal morphologies of Na$_2$SO$_4$·10H$_2$O and Na$_2$HPO$_4$·12H$_2$O show distinct granular and bar shapes, respectively. In Fig. 5(c), the eutectic salt composited Na$_2$SO$_4$·10H$_2$O and Na$_2$HPO$_4$·12H$_2$O can be observed in red circle and rectangle showing granular and bar shapes, respectively. This observation strongly suggests that the EHS is solely a physical mixture of both hydrated salts. Upon the addition of 0.2 wt.% C-2 carbon particles, as depicted in Fig. 5(d), the presence of carbon can be observed within the red circles. Interestingly, the overall morphology of EHS undergoes a significant transformation. EHS with 0.2 wt.% C-2 carbon particles shows lamellar structure

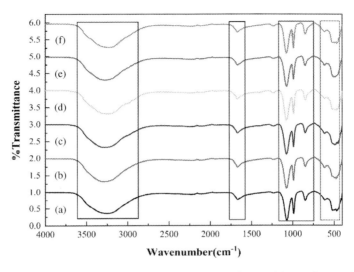

Fig. 6. FTIR spectra of (a) pristine EHS, (b) C-1/EHS, (c) C-2/EHS, (d) C-3/EHS, (e) C-4/EHS and (f) C-5/EHS every composite at 0.2wt% carbon particles.

in red rectangle distinctly. The interaction between carbon particles and EHS will be further analyzed using FTIR in subsequent investigations.

The measured FTIR spectra of pristine EHS and C/EHS composites at 0.2 wt.% carbon particles are shown in Fig. 6. The spectrum of pristine EHS shown as curve (a) had a broad absorption peak in the range of 2750–3650 cm^{-1}, representing hydrogen bonding corresponding to the asymmetric stretching vibration of O-H. A sharp peak at 1668 cm^{-1} corresponds to the bonding vibration of H–O–H. Three characteristic absorption peaks exist at 1071 cm^{-1}, 990 cm^{-1}, and 848 cm^{-1} representing $Na_2HPO_4 \cdot 12H_2O$ and correspond to P–O antisymmetric stretching vibration, P–O stretching vibration, and P–OH antisymmetric stretching vibration, respectively. The peaks at 616 cm^{-1} and 485 cm^{-1} are attributed to the antisymmetric stretching vibration and antisymmetric bending vibration of SO_4^{2-}, corresponding to $Na_2SO_4 \cdot 10H_2O$. The characteristic absorption peaks of $Na_2HPO_4 \cdot 12H_2O$ and $Na_2SO_4 \cdot 10H_2O$ can be seen in the spectra of EHS. The FTIR spectrum of C/EHS composites C-1, C-2, C-3, C-4 and C-5 at 0.2 wt.% are shown in

Figs. 6(b)–6(f), which contain similar characteristic peaks and no other peaks present or peaks shift. This shows that there is no chemical interaction between carbon particles and EHS, and also confirms merely physical interaction between EHS and the addition in EHS.

In conclusion, FTIR and XRD analyses reveal that no chemical interaction is presented between carbon particles and EHS. SEM image analysis indicates a significant change in the crystalline morphology of EHS following the addition of carbon particles. This observation suggests that the presence of carbon particles influences the crystalline morphology of the EHS crystals. The cycling stability and enthalpy of the composite EHS may be impacted.

3.3. *Cycle stability of C/EHS PCMs*

The size of carbon particles plays a crucial role in the stability of NePCM. Micro-nanoparticles possess a higher specific surface area, enabling strong interaction between the particle surface and the surrounding liquid phase. This interaction is scapable of overcoming the density difference between the particles and the liquid, thereby facilitating the stable suspension of particles within the liquid. In the case of EHS PCMs, it undergoes a solid–liquid conversion process, which necessitates not only dispersion stability but also good solid–liquid circulation stability is more important. Figure 7(a) shows pristine EHS as a clarified solution, exhibits uniform dispersion within the EHS even after the addition of 0.2 wt.% carbon particles with different sizes. This observation suggests that the prepared carbon particles possess good dispersibility in EHS. In Fig. 7(b), it is observed that after 200 solid–liquid cycles, the pristine EHS exhibits phase separation. In the case of C-5/EHS, some agglomeration of C-5 particles is observed. The rest of composited PCM does not have phase separation and demonstrates good cycling stability within EHS. In Fig. 7(c), it can be observed that C-1/EHS have agglomerated and no longer dispersed throughout the upper layer after 500 solid–liquid cycles. Similarly, significant agglomeration of C-4/EHS is observed, while in C-5/EHS, the C-5

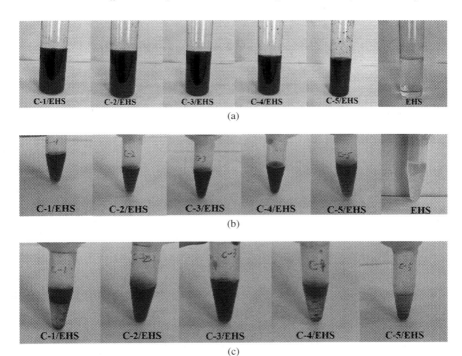

Fig. 7. Digital photographs of sample after 1, 200 and 500 thermal. Cycles (a) Composited C-1/EHS, C-2/EHS, C-3/EHS, C-4/EHS, C-5/EHS with 0.2% mass fraction and pristine EHS before cycle at liquid (from left to right). (b) Composited C-1/EHS, C-2/EHS, C-3/EHS, C-4/EHS, C-5/EHS with 0.2% mass fraction and pristine EHS after 200 cycles at solid (from left to right). (c) Composited C-1/EHS, C-2/EHS, C-3/EHS, C-4/EHS, C-5/EHS with 0.2wt% mass fraction after 500 cycles at solid (from left to right).

carbon particles have completely settled at the bottom. These observations can be attributed to the different sizes (larger or smaller) of the carbon particles, with C-1, C-4, and C-5 carbon particles exhibiting poor cycling stability. In summary, the addition of carbon particles has proven to be beneficial in mitigating the phase separation of EHS. This improvement can be attributed to the high specific surface area of carbon particles, which effectively slows down salt deposition and facilitates complete crystallization within EHS. The C/EHS NePCM exhibits good cycling dispersion stability after 200 cycles of solid–liquid. The C-2/EHS and C-3/EHS PCMs demonstrate sustained stability even after 500 cycles of solid–liquid. These findings suggest that the incorporation of carbon particles enhances the overall stability and performance of the EHS, making them suitable for prolonged cycling applications.

3.4. *Thermal conductivity*

As shown in Fig. 8, the thermal conductivity of pristine EHS was 0.58 W·m^{-1}K^{-1}, and the results of thermal conductivity after adding 0.2 wt.% of carbon particles, respectively, are shown in Table 2. Among the different carbon particles addition, the highest improvement in thermal conductivity was observed in C-1/EHS with a significant increase of 42.1%. This is followed by C-2/EHS, which shows an improvement of 39.3% in thermal conductivity. Results indicate that smaller carbon particle contributes to improved thermal conductivity. This can be attributed to two factors. Smaller size results in a larger specific surface area, allowing for enhanced heat transfer between phase change material and surround. C-1 and C-2 exhibit greater stability in the dispersion within the EHS,

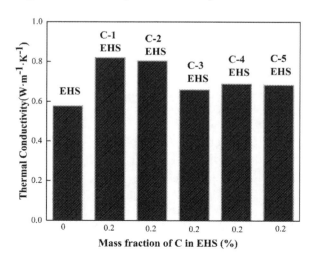

Fig. 8. Thermal conductivity of pristine EHS and C/EHS composites C-1, C-2, C-3, C-4 and C-5 at the 0.2 wt% of carbon particles.

Table 2. Thermal conductivity of pristine EHS and EHS composited at 0.2 wt.% different sized carbon particles.

Sample	Wt.% of carbon particles	Thermal conductivity (W·m^{-1}·g^{-1})	Improvement (%)
Pristine EHS	0	0.58	—
C-1/EHS	0.2	0.82	42.1
C-2/EHS	0.2	0.80	39.9
C-3/EHS	0.2	0.66	14.4
C-4/EHS	0.2	0.69	19.5
C-5/EHS	0.2	0.68	18.8

leading to increased particle-fluid interactions and collisions, thereby facilitating more efficient heat energy transfer within the fluid.[32] However, the performance of supercooling and cycling stability of C-2/EHS is superior compared to C-1/EHS. The thermal conductivity of EHS with the different mass fractions of C-2 carbon particles is further analyzed. As shown in Fig. S4, the thermal conductivity enhancement of EHS initially is profound for the added weight concentration of carbon particles in the range of 0.2 wt.% to 0.8 wt.%. This can be attributed to the reduction of EHS voids by the addition of high thermal conductivity carbon particles. After reaching an addition amount of 0.8 wt.%, excessive carbon particle addition leads to partial deposition and agglomeration. The agglomeration of particles results in the non-uniform composite, which disrupts the formation of an active thermal conducting network within the matrix. The increased concentration of carbon particles forms a high thermal boundary resistance[33] and thus hinders further enhancement of the thermal conductivity of EHS.

3.5. *Thermal performance analysis by DSC*

Figure 9(a) shows the DSC curves of C/EHS composited materials with 0.2% mass fraction. The thermal properties inferred from Fig. 9 are summarized in Table S3. The phase transition temperature

150 Z. Sun et al.

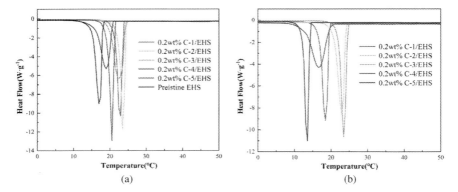

Fig. 9. DSC curve of (a) EHS with different sized carbon particles before cycle (b) EHS with different sized carbon particles after 200 cycles.

of pristine EHS is 18.4°C. Specifically, the transition point increased by 5.5, 6.6, 6.1, 3.0, and 2.5°C, respectively, upon adding 0.2 wt.% of C-1, C-2, C-3, C-4, and C-5 carbon particles. These results confirm that the addition of carbon particles effectively reduces the supercooling degree of EHS. This observation aligns with the findings depicted in the step-cooling curves shown in Fig. S2. Unfortunately, it should be noted that the DSC measurement may introduce slight deviations in the determined transition temperature due to the temperature hysteresis of the instrument. In addition, the phase transition enthalpy of pristine EHS was determined to be 225.8 $J·g^{-1}$. It can be seen that a slight increase in the enthalpy of EHS upon adding 0.2 wt.% of C-2, C-2, C-3, and C-4. This may be because the carbon particles improved the crystallization process of EHS, leading to a more uniform and dense crystalline structure. As a result, the crystallization enthalpy is increased. This is inferred from the dramatic change in C-2/EHS morphology shown in Fig. 5. Figure 9(b) presents the DSC curves of the composites after undergoing 200 solid–liquid cycles. It can be observed that the enthalpy of C/EHS remains relatively stable after 200 solid–liquid cycles, with all values decreasing within a range of 10 $J·g^{-1}$. The phase transition temperature of C-1/EHS and C-5/EHS has decreased by 3.9°C and 5.9°C, respectively. In contrast, the

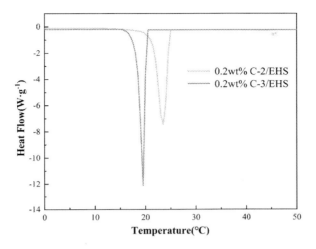

Fig. 10. DSC curve of EHS with C-2 and C-3 after 500 cycles.

phase transition temperature of the other composites shows minimal variation after cycling. This further highlights the poor cycling stability of C-1/EHS and C-5/EHS compared to the other composites resulting in supercooling.

After 500 cycles, C-2/EHS and C-3/EHS maintained good dispersion stability, and their DSC curves were depicted in Fig. 10. Results show that the phase transition temperature of C-2/EHS remains relatively stable, while the phase transition point of C-3/EHS decreases to 20.5°C after 500 solid–liquid cycles. This indicates that C-2/EHS exhibits higher cycling stability. Furthermore, the latent heat of phase change values for C-2/EHS and C-3/EHS, respectively, decreased by 16.9% and 13.3% after 500 cycles. These enthalpy values remain above 200 J·g^{-1}. Therefore, C-2/EHS and C-3/EHS are promising for a wide range of applications.

3.6. *Thermal analysis by TGA*

Figure 11 shows the thermogravimetric curves of pristine EHS and C-2/EHS. The weight loss process of pure EHS can be divided into three stages. In the first stage, 19.7 wt.% of water was lost, followed

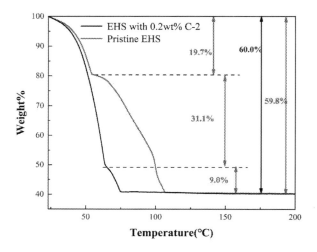

Fig. 11. TGA curves of Pristine EHS and EHS with 0.2wt% C-2.

by the second stage where 31.1 wt.% of water was lost, and an additional 9.0 wt.% of water was lost in the third stage. Pure EHS is composed of a 2:8 ratio of $Na_2SO_4 \cdot 10H_2O$ and $Na_2HPO_4 \cdot 12H_2O$. Based on the water loss process of $Na_2SO_4 \cdot 10H_2O$[34] and $Na_2HPO_4 \cdot 12H_2O$[14], it can be analyzed that the first stage corresponds to the loss of five crystalline water of $Na_2HPO_4 \cdot 12H_2O$, the second stage corresponds to the loss of five crystalline water of $Na_2HPO_4 \cdot 12H_2O$ along with the loss of ten crystalline water, the third stage corresponds to the loss of two crystalline water of $Na_2HPO_4 \cdot 12H_2O$. After the addition of 0.2 wt.% C-2, the water loss process of EHS exhibited only two stages. The first and second stages were combined, resulting in the simultaneous loss of ten crystalline water of $Na_2HPO_4 \cdot 12H_2O$ and ten crystalline water of $Na_2SO_4 \cdot 10H_2O$. The total water loss masses of EHS and C-2/EHS were found to be 59.8% and 60%, respectively, which were in close agreement with the theoretical value of 59.5%. This indicates that the addition of carbon particles had an impact on the water loss process of EHS but did not significantly alter the overall results. Pristine EHS exhibited complete water loss at 107°C, whereas C-2/EHS achieved complete water loss at 75°C. This observation suggests that the addition of C-2 carbon particles indeed accelerates the water loss rate of EHS and enhances its heat transfer efficiency.

4. Conclusion

This paper presents an investigation on the effects of carbon micro-nanoparticles size (123.1 nm, 229.2 nm, 335.1 nm, 456.2 nm and 534.0 nm) on dispersion stability and thermal properties of $Na_2SO_4·10H_2O$–$Na_2HPO_4·12H_2O$ (EHS) phase change materials. The carbon micro-nanoparticles with different size can be uniformly dispersion within EHS, which effectively addresses the issue of phase separation. But the dispersion stability of carbon micro-nanoparticles with different sizes varied clearly following solid–liquid cycling. Carbon particles with sizes of 229.2 nm and 335.1 nm maintained good dispersion stability even after 500 cycles. The addition of carbon particles proved to be effective in reducing the supercooling degree of EHS. Specifically, the supercooling degree decreased to 1.5°C after incorporating 0.2 wt.% of C-2 carbon particles. It is noteworthy that the supercooling degree of EHS increases with the mass fraction of C-2 particles added. The thermal conductivity of C/EHS increased by 42.1%, 39.9%, 14.4%, 19.5% and 18.8% with 0.2 wt.% carbon particles (C-1, C-2, C-3, C-4 and C-5), respectively. As the mass fraction of C-2 increased further, the thermal conductivity of EHS exhibited a trend of initially increasing and then stabilizing. The addition of C-1, C-2, C-3, and C-4 carbon micro-nanoparticles resulted in an improvement in the latent heat of EHS. Specifically, the addition of 0.2 wt.% C-2 increased the latent heat of phase change of EHS to 243.4 J/g. Furthermore, even after 200 cycles of solid–liquid cycling, the latent heat remained above 200 J/g, indicating good thermal stability. The thermogravimetric analysis confirmed the chemical and thermal stability of C-2/EHS. Results demonstrated that the addition of 0.2 wt.% C-2 improved heat transfer efficiency in the EHS phase change material.

Acknowledgment

The authors would like to thank the financial supports from the Natural Science Foundation of Qinghai Province (Grant Nos. 2021-ZJ-906).

ORCID

Zengbao Sun ⊚ https://orcid.org/0009-0001-3934-8958
Xin Liu ⊚ https://orcid.org/0000-0002-5138-4062
Shengnian Tie ⊚ https://orcid.org/0000-0002-1179-4067
Changan Wang ⊚ https://orcid.org/0000-0002-1486-3830

References

1. K. Haruka *et al.*, *Build. Environ.* **238**, 110351 (2023).
2. K. Kr *et al.*, *Buildings* **13**, 79 (2022).
3. C. Yanhu *et al.*, *J. Energy Storage* **67**, 107600 (2023).
4. N. Safna and K. Igor, *Sustainable Energy Technol. Assess.* **52**, (2022) 102241.
5. L. Panpan *et al.*, *ACS nano* **16**, 13715 (2022).
6. Z. Wei *et al.*, *Ind. Eng. Chem Res.* **62**, 7540 (2023).
7. M. Karolina *et al.*, *Chem. Rev.* **123**, 491 (2022).
8. N. Zhang *et al.*, *Adv. Eng. Mater.* **20**, 1700753 (2018).
9. B. Zhao *et al.*, *Energy Fuels* **36**, 10354 (2022).
10. W. Su *et al.*, *Renew. Sustain. Energy Rev.* **48**, 373 (2015).
11. N. Xie *et al.*, *Int. J. Refrig.* **110**, 178 (2020).
12. Y. Liu *et al.*, *Appl. Therm. Eng.* **112**, 208 (2017).
13. D. Xing *et al.*, *Energy Fuels* **36**, 4947 (2022).
14. N. Xie *et al.*, *J. Build. Eng.* **56**, 104747 (2022).
15. A. G. Krishna *et al.*, *Therm. Sci. Eng. Prog.* **28**, 1 (2021).
16. A. Nematpour Keshteli and M. Sheikholeslami, *J. Mol. Liq.* **274**, 516 (2019).
17. J. M. Mahdi S. Lohrasbi and E. C. Nsofor, *Int. J. Heat Mass Transfer* **137**, 630 (2019).
18. Y. B. Tao and Y. -L. He, *Renew. Sustain. Energy Rev.* **93**, 245 (2018).
19. K. W. Shah, *Energy Build.* **175**, 57 (2018).
20. H. Asgharian and E. Baniasadi, *J. Energy Storage* **21**, 186 (2019).

21. Y. Lin *et al.*, *Renewable Sustainable Energy Rev.* **82**, 2730 (2018).
22. X. Liu *et al.*, *J. Mater. Res. Technol.* **12**, 1434 (2021).
23. F. Wang *et al.*, *J. Energy Storage* **53**, 105079 (2022).
24. M. Li and B. Mu, *Appl. Energy* **242**, 695 (2019).
25. K. W. Shah, *Energy Build.* **175**, 57 (2018).
26. M. He *et al.*, *J. Energy Storage* **25**, 100874 (2019).
27. R. R. Kumar *et al.*, *Sol. Energy Mater. Sol. Cells* **240**, 111697 (2022).
28. B Sushrut *et al.*, *ACS Appl. Mater. Interfaces* **9**, 18925 (2017).
29. S. Bertolazzi *et al.*, *ACS Nano* **7**, 3246 (2013).
30. N. Gupta *et al.*, *J. Energy Storage* **41**, 102899 (2021).
31. S. Kannan *et al.*, *Ind. Eng. Chem. Res.* **61**, 16341 (2022).
32. M. Chopkar *et al.*, *Scr. Mater.* **55**, 1017 (2006).
33. O. Mahian *et al.*, *Phys. Rep.* **790**, 1 (2019).
34. H. E. Brand *et al.*, *Phys. Chem. Miner.* **36**(1), 29 (2008).

Chapter 9

An efficient CoSe$_2$-Co$_3$O$_4$-Ag hybrid catalyst for electrocatalytic oxygen evolution

Qichen Liang, Nana Du * and Huajie Xu

School of Chemical and Material Engineering
Fuyang Normal University, Fuyang 236037, P. R. China
*dunana@fynu.edu.cn; fysfxybylw@163.com

The development of cheap, high activity and stability catalysts for oxygen evolution reaction (OER) is of great significance because of its important role in energy storage technology. Herein, flower-like CoSe$_2$-Co$_3$O$_4$-Ag with enhanced electrochemical performance was synthesized by simple methods. The CoSe$_2$-Co$_3$O$_4$-Ag exhibits lower overpotential and higher current density, with an overpotential of 324 mV at the current density of 10 mA·cm^{-2} and a Tafel slope of 65.69 mA·dec^{-1}. From the ECSA normalized LSV curves, higher electrocatalytic activity for OER is mainly due to the increase of ESCA and conductivity, minor due to adjusting the electronic configuration. These strategies to improve electrochemical active area and conductivity have valuable reference for designing cheap, high activity and stability catalysts.

Keywords: Oxygen evolution reaction; CoSe$_2$-Co$_3$O$_4$-Ag; electrochemical active surface area; conductivity.

1. Introduction

Oxygen evolution reaction (OER) is a key reaction of energy storage technologies, such as hydrogen production from water splitting and metal air batteries. It has received growing attention because

---

*Corresponding author.
To cite this article, please refer to its earlier version published in the Functional Materials Letters, Volume 16(8), 2340033 (2023), DOI: 10.1142/S1793604723400337.

of the important role of energy storage technologies in coping with energy crises and environmental problems.[1-4] However, OER has a high overpotential because of involving a complex four-electron reaction process, so effective catalysts can be designed to facilitate the sluggish kinetics of OER.[5,6] Currently, noble metal oxides such as IrO_2 and RuO_2 generally possess outstanding activity for OER, but their low abundance and high cost seriously impede their large-scale application.[7-9] Therefore, seeking efficient, stable, inexpensive catalysts that can be applied on a large scale is crucial for these energy storage technologies.[10-12]

Transition metal compounds are terrific substitutes for noble metal catalysts due to their abundant reserves and potential high efficiency, which have attracted more and more attention.[13] Currently, transition metal compounds such as metal hydroxides,[14] nitrides,[15] sulfides[16] and borides[17] were developed to effectively catalyze OER. Thereinto, cubic $CoSe_2$ is considered as a promising OER electrocatalyst because its electronic structure is similar to the ideal e_g filling proposed by the Yang Shao-Horn principle.[18-20] Some measures have been proposed to optimize the electrocatalytic performance of $CoSe_2$, such as adjusting the electronic configuration,[21] improving the conductivity and increasing the number of active sites,[22] etc. These studies are of great significance for obtaining highly active OER catalysts. For instance, the electrocatalytic performance of developed hybrids of various metal or metal oxide nanoparticles (NPs) with $CoSe_2$ exhibit much better OER performance than the original $CoSe_2$ due to the electron coupling effects.[23-25] The synergistic effect of the two substances adjusts the electronic configuration and leads to the improvement of electrocatalytic activity. Based on the above understanding, it is natural to develop new OER catalysts by hybridizing metal oxide with $CoSe_2$. Co_3O_4 is an active OER catalyst that has been extensively studied. Thus, the combination of Co_3O_4 and $CoSe_2$ is expected to obtain high performance OER catalyst. Herein, we report a facile *in situ* synthesis strategy for the formation of Co_3O_4 NPs on the flower-like $CoSe_2$ by adding ammonia.

In addition, high conductivity increases the rate of electron transfer and thus reduces the Schottky barrier at the interfaces, which is also a prerequisite for excellent electrocatalytic performance of catalysts. Therefore, it is of great significance to improve the conductivity of developed catalysts. It has been reported that the improvement in the conductivity of metallic compounds by cation exchange is due to the growing mobility of metal cations in the anionic framework after the metal cations have replaced the original cations under moderate reaction conditions. For instance, $Ni_xCo_{(3-x)}O_4$/NH_2-CNTs was prepared by cation exchange reaction with enhanced bifunctional electrocatalytic activity towards ORR and OER.[26] Here, the doped nickel ions increase the electron migration rate during the reaction to achieve an increase in electrocatalytic activity. Herein, $CoSe_2$-Co_3O_4-Ag with enhanced conductivity was prepared by cation exchange with a trace amount of silver ions. The systematic study of OER activity showed that the as-prepared $CoSe_2$-Co_3O_4-Ag possesses relatively higher electrocatalytic activity for OER. The OER overpotential of $CoSe_2$-Co_3O_4-Ag was 324 mV at the current density of 10 mA·cm^{-2}, which was much lower than that of $CoSe_2$, Co_3O_4 and $CoSe_2$-Co_3O_4. Moreover, $CoSe_2$-Co_3O_4-Ag has the highest current density (24.68 mA·cm^{-2}) at 0.35 V overpotential, and the lowest Tafel slope (65.69 mA·dec^{-1}). The ECSA normalized current density of $CoSe_2$-Co_3O_4-Ag is also larger than that of other catalysts due to the enhancement of intrinsic activity.

2. Experimental Section

2.1. Synthesis of flower-like Co(OH)$_2$

The flower-like Co(OH)$_2$ was prepared according to the previous report.[27] Typically, 10.91362 g Co(NO$_3$)$_2$·6H$_2$O was dissolved in 30 ml anhydrous methanol solution as A solution, and 2.463 g 2-methylimidazole was dissolved in 12 ml anhydrous methanol solution as B solution. When the solid was fully dissolved, B was dissolved in A and mixed at room temperature for 10 min. Then, the mixed solution was ultrasonic at room temperature for 30 min. The precipitation was collected by centrifugation, washed three times with deionized water and anhydrous ethanol, and dried by vacuum at 60°C for 10 h.

2.2. *Synthesis of flower-like* $CoSe_2$

Ultrasonic dispersion of 1mmol of synthetic cobalt oxide in 50 ml deionized water was added into a 100 ml three-way flask and heated to 50°C. Argon gas was injected to remove the air from the solution, and then 2 mmol selenium powder was added and stirred for 10 min. Then, 10 ml of water solution containing 4 mmol KBH_4 was added to the mixed solution within 15 min. The black solution after the reaction was placed in 100 ml Teflon-lined autoclave, and reacted in an oven at 180°C for 15 h. After the autoclave had been cooled to room temperature, the black precipitate was collected by centrifugation. The product was washed twice with deionized water and anhydrous ethanol, respectively, and dried at 60°C for 10 h.

2.3. *Synthesis of* $CoSe_2$-Co_3O_4

The $CoSe_2$ powder of the previous step was dispersed into 25 ml anhydrous ethanol, and 40, 60, 80, 100, and 120 µL ammonia were added, respectively. The powder was magnetically stirred at room temperature for 10 min. Then, the black solution was placed in a 50 ml Teflon-lined autoclave, and finally reacted in an oven at 150°C for 3 h. After the autoclave had been cooled to room temperature, the black precipitate was collected by centrifugation, washed twice with deionized water and anhydrous ethanol, respectively, and dried at 60°C for 10 h under vacuum. In this paper, $CoSe_2$-Co_3O_4 refers to the product when the amount of ammonia is 100 µL.

2.4. *Synthesis of* $CoSe_2$-Co_3O_4-*Ag*

The powder of $CoSe_2$-Co_3O_4 was dispersed in 30 ml deionized water and added with 5 mg $AgNO_3$. After stirring magnetically at room temperature for 30 min, the powder was placed in 50 ml Teflon-lined autoclave at 160°C for 6 h. The powder was washed three times with ethanol and deionized water to obtain $CoSe_2$-Co_3O_4-Ag.

2.5. Synthesis of Co_3O_4

Weigh 0.5 g $Co(NO_3)_2 \cdot 6H_2O$ and dissolve it in 25 ml ethanol, then add 2.5 ml ammonia water and stirring for 10 min. The solution was heated in a 30 ml Teflon-lined autoclave at 150°C for 3 h. The powder was washed three times with ethanol and deionized water to obtain Co_3O_4.

2.6. Electrochemical measurement

All electrochemical measurements were carried out at the electrochemistry station (CHI 760E, Shanghai Chenhua, China) using a three-electrode system in 1 M KOH electrolyte. The working electrode is a glassy carbon rotating disk electrode (GC RDE, PINE, PA, USA) with a geometric area of 0.19625 cm^2. Saturated calomel electrode (SCE) and platinum foil electrode were used as reference electrode and counter electrode, respectively. In the work, all potentials were converted to reversible hydrogen electrodes (RHE). In 1 M KOH, $E_{RHE} = E_{SCE} + 1.07$ V.

5 mg catalyst powder was dispersed into a mixture containing 0.5 ml deionized water, 0.5 ml N–N dimethylformamide and 40 μL (5% wt%) Nafion solution to form a uniform solution by ultrasonic. Then, 8 μL mixed Solution was transferred to the surface of the glassy carbon rotating disk electrode. After drying in air at room temperature, catalyst film was formed on the electrode surface. All catalyst loads were 0.2 mg·cm^{-2}.

The electrochemical measurements were performed in an oxygen-saturated 1 M KOH electrolyte, which bubbled for at least 30 min. During the OER experiment, oxygen was continuously maintained above the electrolyte to ensure that the electrolyte was oxygen saturated at all times. The linear scanning voltammetry (LCV) was carried out ranging from 1.2 V to 1.8 V (versus RHE) with a sweep rate of 5 mV·s^{-1} and a rotating speed of 1600 rpm. The accelerated stability tests were carried out 2000 cycles between 1.2 V and 1.8 V (versus RHE) potential at a scan rate of 100 mV·s^{-1}. Then, after that, the OER polarization data were recorded at a scan rate of 5 mV·s^{-1}. All the polarization data were corrected by IR-compensation to reduce the impacts from the solution resistance. Electrical impedance spectroscopy (EIS) was measured in the same configuration with frequencies ranging from 100 kHz to 1 KHz. Cyclic voltammetry (CV) curves were obtained by potential cycling (versus RHE) at a scan rate of 5 mV·s^{-1} in an argon-saturated electrolyte. Two-layer capacitance (C_{dl}) was obtained from the CV curves between 1.32 V and 1.42 V (versus RHE) by $C_{dl} = I/\nu$ at 20, 40, 60, 80 and 100 mV·s^{-1}, respectively, where I is the difference of charging current density ($\Delta j/2 = j_a - j_c$) at 1.37 V (versus RHE), and ν is the scan rate.

Tafel plots were obtained by plotting overpotential (η) versus log current (log j) according to the Tafel equation ($\eta = b \log(j) + a$). The Tafel slope as a quantitative indicator of the electrocatalytic activity of the catalyst was determined by fitting the Tafel plots.

3. Results and Discussions

CoSe$_2$-Co$_3$O$_4$-Ag was synthesized in the following steps. To begin with, the flower-like cobalt hydroxide was synthesized as template by a facile method. Then, flower-like CoSe$_2$ was successfully obtained through selenization reaction using selenium powder as selenium source, and its morphology remained basically unchanged after reaction, as shown in Figs. S1(a) and S1(b) in the supporting information. After that, a small amount of ammonia was added to form Co$_3$O$_4$ particles on the surface of CoSe$_2$. added ammonia combined with free cobalt ions in cobalt selenide to form [Co(NH$_3$)$_6$]$^{2+}$, which promoted the dissolution of CoSe$_2$ to further form more [Co(NH$_3$)$_6$]$^{2+}$. Part of [Co(NH$_3$)$_6$]$^{2+}$ ions were oxidized to

$[Co(NH_3)_6]^{3+}$ by oxygen in the air, and then decomposed to Co_3O_4 at a certain temperature.[28] Co_3O_4 particles with tens of nanometers can be clearly seen on the flower-like surface of $CoSe_2$, as shown in Figs. S1(c) and S1(d).

Moreover, the surface of product was coarser after ammonia treatment by comparing the morphology of $CoSe_2$-Co_3O_4 and $CoSe_2$ (Fig. S2). Finally, Ag^+ was doped into $CoSe_2$-Co_3O_4 by ion exchange in the reactor. The TEM and SEM images of $CoSe_2$-Co_3O_4-Ag are shown in Figs. 1(a) and 1(b). Co_3O_4 NPs with the size of 30–80 nm were evenly dispersed on the flower-like $CoSe_2$, without obvious particle aggregation. Compared with the morphology of $CoSe_2$-Co_3O_4 and $CoSe_2$-Co_3O_4-Ag, it can be seen that the introduction of Ag^+ had no effect on the morphology of the product. Figure 1(c) shows the high-resolution transmission electron microscopy (HRTEM) image of $CoSe_2$-Co_3O_4-Ag. The image revealed that the

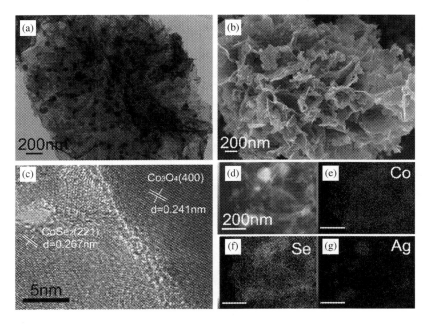

Fig. 1. (a) TEM, (b) SEM and (c) HRTEM images of flower-like $CoSe_2$-Co_3O_4-Ag, (d)–(g) HAADF-STEM-EDS mapping of $CoSe_2$-Co_3O_4-Ag.

lattice spacing of Co_3O_4 is 0.241 nm, corresponding to (400) crystal plane of cubic phase Co_3O_4,[28] and the lattice spacing of $CoSe_2$ 0.267 nm, corresponding to (221) crystal planes of cubic phase $CoSe_2$.[29] Energy dispersion spectrum (EDS) element mapping image (Figs. 1(d)–1(g) and Fig. S3) proved the existence of Co, Se and Ag. Noteworthily, the distribution of Ag elements showed that Ag doping mainly concentrates on small particles of Co_3O_4. Conclusively, the distribution of elements can further confirm the successful synthesis of $CoSe_2$-Co_3O_4-Ag.

As shown in Fig. 2(a), X-ray diffraction (XRD) was used to analyze the crystal patterns of the product. The XRD peaks at 34.2°, 46.4°, 51.7° and 63.4° correspond to (210), (221), (311) and (400) crystal planes of cubic phase $CoSe_2$ (JCPDS 09–0234),[29] and the XRD peaks at 31.2°, 36.8°, 44.8°, 65.2° correspond to (220), (311), (400), (440) crystal planes of cubic phase Co_3O_4 (JCPDS 42–1467).[28] It should be noted that the XRD peak of $CoSe_2$ (210) in $CoSe_2$-Co_3O_4 shifted to higher angle compared with that in pure $CoSe_2$. The result is consistent with the HRTEM image in Fig. 1(c). Furthermore, it can also be seen that the angle of the XRD peak of $CoSe_2$ (210) increased correspondingly with the increase of ammonia content (Fig. S4). These results indicated that the crystal structure of $CoSe_2$-Co_3O_4 is changed by the introduction of Co_3O_4 to some extent. Correspondingly, the XRD pattern of $CoSe_2$-Co_3O_4-Ag showed no significant change compared with that of $CoSe_2$-Co_3O_4, indicating that the crystal structure change can be ignored after the introduction of Ag^+. Figure 2(b) shows the Raman spectra of $CoSe_2$, $CoSe_2$-Co_3O_4, and $CoSe_2$-Co_3O_4-Ag. The load of Co_3O_4 causes the peak of the original $CoSe_2$ (681 cm^{-1}) to move to the left to the peak (676 cm^{-1}) due to the interaction of the Co_3O_4 with $CoSe_2$. However, Ag^+ doping hardly changed the position of the peak because that Ag doping mainly concentrates on the Co_3O_4.

In order to further analyze the electronic structures of $CoSe_2$, $CoSe_2$-Co_3O_4 and $CoSe_2$-Co_3O_4-Ag, X-ray photoelectron spectroscopy (XPS) was used to further analyze the materials, as shown in Fig. 2(c). The X-ray photoelectron spectra of $CoSe_2$-Co_3O_4-Ag suggest the presence of Co, Se, Ag and O, which is consistent with the

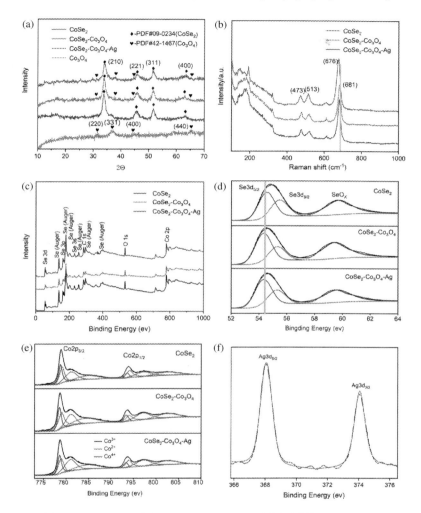

Fig. 2. (a) XRD pattern, (b) Raman spectra and (c) XPS spectra of the $CoSe_2$, $CoSe_2$-Co_3O_4 and $CoSe_2$-Co_3O_4-Ag, XPS spectra of the (d) Se $3d$, (e) Co $2d$ and (f) Ag $3d$.

literature.[30] Figure 2(d) shows the comparison of Se$3d$ spectra of $CoSe_2$, $CoSe_2$-Co_3O_4 and $CoSe_2$-Co_3O_4-Ag. The peak of binding energy at 54.35 and 55.49 eV belongs to the Se $3d_{3/2}$ peak and Se $3d_{5/2}$ of Se_2^{2-} in the $CoSe_2$, respectively, while the peak of binding energy at 59.8 eV in the spectra is caused by SeO_x.[31] In contrast, the SeO$_x$ peak of CoSe$_2$-Co$_3$O$_4$ is shifted to 59.5 eV. However, the peaks of CoSe$_2$-Co$_3$O$_4$-Ag did not shift, indicating that silver doping had little effect on the electronic structure of the material. The peak around 778.7 eV in Co 2p spectral of CoSe$_2$ can be decomposed into three peaks at 778.6, 779 and 781.1 eV, respectively, corresponding to Co^{3+}, Co^{2+} and Co^{4+},[31] as shown in Fig. 2(e). The peak at 784.6 eV is the accompanying peak of Co2$p_{3/2}$. Likewise, the peak of Co2$p_{1/2}$ at 793.6 eV was decomposed into corresponding peaks of Co^{3+}, Co^{2+}, Co^{4+}. The peak at 793.5, 794.2 and 797.4 eV is the accompanying peak of Co2$p_{1/2}$. The peak strength of Co^{4+} increased to a certain extent after Co$_3$O$_4$ particles were loaded on the surface of CoSe$_2$ by comparing the Co2p spectra of samples. It has been reported that Co^{4+} is the real active site when co-based catalysts are used to catalyze OER. Therefore, loading Co$_3$O$_4$ particles on the surface of CoSe$_2$ increases the active site, which is expected to improve the catalytic activity of OER. Moreover, the binding energy of Co and Se elements in CoSe$_2$-Co$_3$O$_4$ and CoSe$_2$-Co$_3$O$_4$-Ag was lower than that of CoSe$_2$ (Figs. 2(d) and 2(e)), indicating that the loading of Co$_3$O$_4$ changed the electronic structure. In conclusion, the loading of Co$_3$O$_4$ NPs on CoSe$_2$ reduces the binding energy of elements, changes the electronic structure and forms more active sites. The spectra of Ag 3d were attributed to Ag$^+$ in CoSe$_2$-Co$_3$O$_4$-Ag,[32] confirming the successful exchange of Ag$^+$ to the surface of CoSe$_2$-Co$_3$O$_4$ by ion exchange cation, which is consistent with the results in Fig. 1(f).

The OER catalytic activity of catalysts was studied in oxygen-saturated 1M KOH electrolyte using a three-electrode system. Under the same conditions, the electrochemical tests were performed on CoSe$_2$, Co$_3$O$_4$, CoSe$_2$-Co$_3$O$_4$ with different Co$_3$O$_4$ molar ratios, CoSe$_2$-Co$_3$O$_4$-Ag and commercial RuO$_2$ for comparison. The CV curve of catalysts is shown in Fig. S5(a). CoSe$_2$, Co$_3$O$_4$, CoSe$_2$-Co$_3$O$_4$ and CoSe$_2$-Co$_3$O$_4$-Ag have similar CV curves, and the redox peaks are located at 1.51–1.61 belongs to the redox peak of Co^{2+}/Co^{3+}. The position of the redox peak is consistent with the literature, which accords with the characteristics of cobalt-based catalyst.[33] Figure 3(a) shows the LSV curve corrected by IR-compensation

An efficient $CoSe_2$-Co_3O_4-Ag hybrid catalyst 167

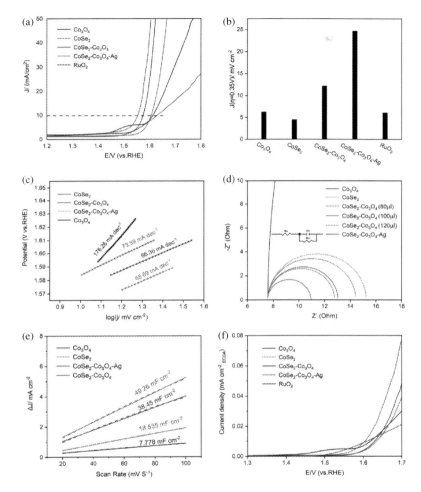

Fig. 3. (a) LSV curves, (b) current densities at an overpotential of 0.35 V, (c) Tafel plots, (d) Nyquist plots (Inset: the equivalent circuit), (e) the double layer capacitance and (f) ECSA-normalized LSV curves.

at a sweep speed of 5 mV·s^{-1}. Obviously, the polarization curve of $CoSe_2$-Co_3O_4-Ag has earlier onset potential and higher current density. As one of the important indexes to evaluate the electrocatalytic activity of OER, the overpotential of $CoSe_2$-Co_3O_4-Ag is the lowest, which is 324 mV at the current density of 10 mA·cm^{-2}. In contrast, the overpotential of $CoSe_2$, Co_3O_4, $CoSe_2$-Co_3O_4 and commercial RuO_2 at the current density of 10 mA·cm^{-2} is 376, 383, 345 and 400 mV, respectively. The current densities of all the catalysts at a fixed overpotential of 350 mV are shown in Fig. 3(b) and Fig. S5(b), to visually compare the electrocatalytic performance. The current density of $CoSe_2$-Co_3O_4-Ag is the highest, reaching 24.68 mA·cm^{-2}, which is 2.03 times that of $CoSe_2$-Co_3O_4, 5.48 times that of $CoSe_2$ and 3.87 times that of Co_3O_4, respectively. The current density of $CoSe_2$-Co_3O_4 with different amounts of ammonia increases at the beginning and then decreases with the increase of the amount of ammonia (Fig. S5(b)). Under the same voltage, the higher the current density, the more favorable the catalytic performance of the material.

Tafel slopes of various catalysts were determined to estimate the OER kinetics, as shown in Fig. 3(c). The Tafel slope of $CoSe_2$-Co_3O_4-Ag is 65.69 mA·dec^{-1}, lower than that of $CoSe_2$, Co_3O_4 and $CoSe_2$-Co_3O_4, indicating the more rapid OER rate owing to the action of Co_3O_4 and the introduction of Ag^+. Furthermore, the electrochemical impedance spectroscopy (EIS) of OER catalysts with different compositions were tested to further estimate the OER kinetics (Fig. 3(d)). The plots could be fitted by a solution resistor (R_s) in series with a parallel unit of a charge transfer resistor (R_{ct}) and a constant phase element (CPE) (equivalent circuit, Fig. 3(d) inset; fitting parameters are shown in Table S1). $CoSe_2$-Co_3O_4-Ag possessed small charge transfer resistance of 3.374 Ω (Table S1), which is smaller than the value of other catalysts. The smaller charge transfer resistance indicates faster mass transfer kinetics for OER and higher efficient electrochemical properties, which is strongly due to the fast electron transfer rate caused by the introduction Co_3O_4 and of Ag^+. The Nyquist plots at different electrode potentials were collected to further understand the reaction mechanism, the results are shown in Figs. S6(a)–S6(c). Using the Bode plot, the plots of phase angle versus log(frequency) at 1.3–1.6 V (versus RHE) were obtained (Figs. S6(d)–S6(f)). In Fig. S6, the semicircle of Nyquist plot of the three catalysts is reduced continuously with the increase of potential, and the phase peak shifts to a higher frequency. It is shown that none of the three catalysts had signs of the degeneration of the reaction, indicating that the observed current density in the experiment is from OER.[34]

Double-layer capacitance (C_{dl}) is proportional to electrochemical active surface area (ECSA), which was obtained from CV curves between 1.32 V and 1.42 V (versus RHE) at different scanning rates (Figs. S7(c)–S7(f)).[35] As shown in Fig. 3(e), the double-layer capacitance of $CoSe_2$-Co_3O_4 is 49.26 mF·cm^{-2}, much higher than that of $CoSe_2$, indicating that the formation of Co_3O_4 improves the ECSA. Obviously, the loading of Co_3O_4 particles on the surface of $CoSe_2$ increases the specific surface area, thus increasing the active sites and electrochemical active area of the catalyst. The effect of adding different volumes of ammonia on the double layer capacitance was showed in Fig. S7(a). The experimental results show that the double layer capacitance (C_{dl}) increases at the beginning and then decreases with the increase of the amount of ammonia. The maximum C_{dl} can reach 49.26 mF·cm^{-2} when the amount of ammonia is 100 μL. The variation trend of C_{dl} is the same as that of current density (Figs. S5(b) and S7(a)). The reason for this is that the formation of Co_3O_4 is accompanied by the dissolution of $CoSe_2$. At the beginning, the Co_3O_4 particles formed increased correspondingly with the increase of the amount of ammonia, and the increased active sites were more than that lost by the dissolution of $CoSe_2$. However, with the further increase of ammonia addition, the active sites increased by Co_3O_4 particles were less than those lost by dissolved $CoSe_2$. Figure 1 confirmed that the addition of ammonia does etch $CoSe_2$ and form Co_3O_4 particles, which can expose more active sites. Appreciably, the double-layer capacitance of $CoSe_2$-Co_3O_4-Ag is reduced, because the original active site is occupied due to a fraction of Co^{2+} being replaced by Ag^+ through ion exchange. Although the introduction of Ag^+ leads to the reduction of active sites, considering that it can improve the conductivity of the catalyst and increase the OER rate, it is practicable to improve the catalytic activity of the catalyst. To evaluate the intrinsic OER activity of the catalysts, we calculated the ECSA normalized current density by dividing the OER current with ESCA (Figs. 3(f) and S7(b)).[36] Noteworthily, the ECSA normalized current density of CoSe$_2$-Co$_3$O$_4$-Ag is larger than that of CoSe$_2$-Co$_3$O$_4$, from the normalized LSV curves of ECSA (Fig. 3(f)), indicating the higher OER activity of CoSe$_2$-Co$_3$O$_4$-Ag is due to the enhancement of intrinsic activity caused by introduction of silver ions. However, the ECSA normalized current density of CoSe$_2$-Co$_3$O$_4$ is only slightly higher than that of CoSe$_2$, indicating the enhanced OER activity of CoSe$_2$-Co$_3$O$_4$ is mainly due to the increase of ESCA, minor due to the regulating electronic structure caused by synergistic effect of CoSe$_2$ and Co$_3$O$_4$. The ECSA normalized current density is little different among CoSe$_2$-Co$_3$O$_4$ with different amounts of ammonia in Fig. S7(b). Conclusively, the change of catalytic activity of CoSe$_2$-Co$_3$O$_4$ with different amounts of ammonia is due to the change of ESCA.

In conclusion, CoSe$_2$-Co$_3$O$_4$-Ag as a promising OER catalyst was obtained by growing Co$_3$O$_4$ NPs on the CoSe$_2$ surface and introduction of Ag$^+$. Systematic study of electrochemical activity show that the increase of ESCA and the increase of conductivity are the main reasons for the improvement of electrocatalytic activity for OER.

Continuous potential cycling was carried out in 1M KOH solution saturated with O$_2$ in order to evaluate the stability of the catalysts. It can be clearly seen in Fig. 4(a) that the polarization curve

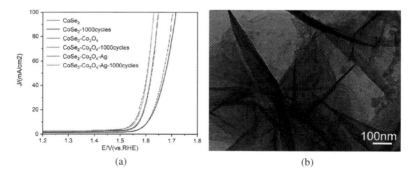

Fig. 4. (a) OER polarization curves of catalysts before and after potential sweeps of 1000 cycles, (b) TEM image of CoSe$_2$-Co$_3$O$_4$-Ag after the durability OER test.

of $CoSe_2$-Co_3O_4-Ag hardly changed after 1000 cycles, indicating that the catalyst has good stability in the OER process. However, the stability of $CoSe_2$-Co_3O_4 and $CoSe_2$-Co_3O_4-Ag is better than that of $CoSe_2$, which can also be confirmed from the figure. TEM image of the tested catalyst confirmed its stability. The morphology of the catalyst was still maintained the original morphology without change after 1000 potential cycles, as shown in Fig. 4(b).

4. Conclusion

In summary, we successfully synthesized the flower-like $CoSe_2$-Co_3O_4-Ag for catalyzing OER in alkaline solution through *in situ* growing Co_3O_4 NPs onto $CoSe_2$ and doping a small amount of Ag^+. The $CoSe_2$-Co_3O_4-Ag possesses relatively high electrocatalytic activity for OER with overpotential of 324 mV at the current density of 10 $mA \cdot cm^{-2}$, which was much lower than that of $CoSe_2$, Co_3O_4 and $CoSe_2$-Co_3O_4. $CoSe_2$-Co_3O_4-Ag exhibits higher electrocatalytic activity for OER mainly due to the increase of ESCA caused by growing Co_3O_4 NPs on the $CoSe_2$ surface and the increase of conductivity caused by introduction of silver ions, minor due to the adjusting electronic configuration caused by the synergistic effect of $CoSe_2$ and Co_3O_4. These results indicate that flower-like $CoSe_2$-Co_3O_4-Ag has good potential for energy conversion and energy storage, which is of great significance for designing transition metal catalysts with high electrochemical active area and appropriate conductivity.

Acknowledgments

This work was supported by the National Natural Science Foundation of China (No. 21701026), the Natural Science Foundation of Anhui Province (No. 1808085MB46), the Key project of Natural Science Research of Anhui Higher Education Institutions (No. KJ2019A0526) and Cooperative project between Fuyang Municipal Government and Fuyang Normal University (No. XDHX201701).

ORCID

Qichen Liang ⓘ https://orcid.org/0009-0009-0255-5003
Nana Du ⓘ https://orcid.org/0000-0001-9676-1215
Huajie Xu ⓘ https://orcid.org/0000-0001-8772-9520

References

1. K. Dang et al., *Nano Res.* **14**, 4848 (2021).
2. J. Zhao et al., *Small* **16**, e2003916 (2020).
3. S. Zuo et al., *Adv. Energy Mater.* **12**, 2103383 (2022).
4. F.-T. Tsai et al., *J. Mater. Chem. A* **8**, 9939 (2020).
5. S. Niu et al., *Appl. Catal. B: Environ.* **297**, 120442 (2021).
6. L. Liang et al., *Nano Energy* **88**, 106221 (2021).
7. C. Ma et al., *ACS Appl. Mater. Interfaces* **12**, 34980 (2020).
8. Y. Wu et al., *ACS Appl. Energy Mater.* **2**, 4105 (2019).
9. T. D. Nguyen et al., *Int. J. Hydrog. Energy* **45**, 46 (2020).
10. M. Lin et al., *Nano Res.* **16**, 2094 (2022).
11. J. Masud et al., *J. Mater. Chem. A* **4**, 9750 (2016).
12. X. Wu et al., *Nano Res.* **13**, 2130 (2020).
13. T. Guo, L. Li and Z. Wang, *Adv. Energy Mater.* **12**, 2200827 (2022).
14. Z. Li et al., *Adv. Sci.* **8**, 2002631 (2021).
15. Y. Guo et al., *Int. J. Hydrog. Energy* **46**, 22268 (2021).
16. H.-J. Liu et al., *Nano Res.* **16**, 5929 (2023).
17. T. Li et al., *J. Mater. Chem. A* **9**, 12283 (2021).
18. S. Prabhakaran et al., *Small* **16**, 2000797 (2020).
19. M. Ramadoss et al., *J. Mater. Sci. Technol.* **78**, 229 (2021).
20. J. Song et al., *Adv. Sci.* **9**, 2104522 (2022).
21. J. Du et al., *J. ACS Appl. Mater. Interfaces* **12**, 686 (2020).
22. F. Razmjooei et al., *ACS Catal.* **7**, 2381 (2017).
23. J. Hu et al., *ACS Catal.* **9**, 10705 (2019).
24. J. Du, F. Li and L. Sun, *Chem. Soc. Rev.* **50**, 2663 (2021).
25. H. Yang et al., *Nano Res.* **9**, 207 (2016).
26. B. Chen et al., *J. Mater. Chem. A* **6**, 9517 (2018).
27. T.-J. Wang et al., *Nano Research* **13**, 79 (2019).
28. Y. Dong et al., *Nanotechnology* **18**, 435602 (2017).
29. H. Yu et al., *Advanced Materials Interfaces* **8**, 2001310 (2020).
30. D. Chen et al., *J. Mater. Chem. A* **8**, 12035 (2020).

31. T. Zhang et al., *Electrochim. Acta* **356**, 136822 (2020).
32. X. L. Zhang et al., *Angew. Chem. Int. Ed.* **60**, 6553 (2021).
33. S. Wan et al., *ACS Sustain. Chem. Eng.* **6**, 15374 (2018).
34. S. Hirai et al., *J. Mater. Chem. A* **7**, 15387 (2019).
35. S. Hirai et al., *RSC Adv.* **12**, 24427 (2022).
36. S. L. Zhang et al., *Adv. Mater.* **32**, 2002235 (2020).

Chapter 10

Simulation and fabrication of titanium dioxide thin films for supercapacitor electrode applications

S. Harish [*], Muhammad Hamza [†], P. Uma Sathyakam [*,§]
and Annamalai Senthil Kumar [‡]

[*]*School of Electrical Engineering, Vellore Institute of Technology*
Vellore 632014, India
[†]*Beijing Institute of Technology*
Beijing 100081, P. R. China
[‡]*CO2 and Green Technologies Research Centre*
Vellore Institute of Technology, Vellore 632014, India
[§]*umasathyakam.p@vit.ac.in*

Nanostructured thin-film electrode materials are proposed for supercapacitors due to their outstanding performance over bulk materials. In this work, we fabricated a TiO_2 nanotube film over a titanium foil using a top-down approach for supercapacitor electrodes. We noticed that the fabricated nanotubes are uniform and well aligned, confirmed by FESEM; the TiO_2 nanotube parameters were further simulated using COMSOL Multiphysics. Simulations show an areal capacitance of 1.19393 pF/cm^2 with oxidation and reduction peak currents of 6.18921 × 10^{-15} A and –6.0320 × 10^{-15} A, respectively, at 10 mV/s scan rate. The as-prepared nanotubes show a poor areal capacitance of 1.0193 F/cm^2, which is improved to 12.8764 F/cm^2 at a scan rate of 10 mV/s, that is approximately 12.63 times with oxidation and reduction peak currents of 0.129 mA/cm^2 and –0.105 mA/cm^2, respectively, by performing an electrochemical etching. Further, the surface roughness of both as-prepared and etched samples is studied to comment on their surface area changes.

[§]Corresponding author.
To cite this article, please refer to its earlier version published in the Functional Materials Letters, Volume 16(8), 2340034 (2023), DOI: 10.1142/S1793604723400349.

176   S. Harish et al.

The effect of the etched sample is studied, compared and validated with simulation, which reveals that the etched TiO$_2$ nanotubes thin-film sample shows considerable similarity with the simulation results.

Keywords: Nanotubes; TiO$_2$; supercapacitor electrode; COMSOL; cyclic voltammetry.

1. Introduction

The advent of electrochemical energy storage is tremendously increasing nowadays.[1] Likewise, technology makes the usage of devices like supercapacitors (SCs) and rechargeable batteries (RBs). RBs like Lithium-Ion batteries (LIBs) are the most commonly used commercial batteries due to their faster charge transfer, greater capacity and energy densities.[2-6] Among these energy storage technologies, SCs are currently trending due to their high-power density, which is considerable than traditional capacitors and also due to fast charge and discharge time, making them desirable in applications like power tools, smart wearable electronics, hybrid energy storage systems in EVs and so on.[7-10]

SCs are classified into three types based on their charge storage types: Electrochemical double layer capacitance (EDLC), pseudo and hybrid capacitors, where the charge storage happens electrostatically, electrochemically, and both electrostatically and chemically, in these devices, respectively.[11]

Titanium dioxide (TiO$_2$) is a commonly used electrode material in energy harvesting like solar cells,[12] photoelectrochemical cells (PEC),[13] energy storage-like RBs,[14] SCs,[15] and bio-medical applications.[16,17]

TiO$_2$ is structurally, physically, and chemically available in three phases: anatase, rutile, and brookite. Their properties, like non-reactive, non-toxic, environment-friendly, easy availability, and cost-effectiveness, make it a good choice.[18] Apart from all those properties, TiO$_2$ is more stable. Each phase has a different crystalline structure, and bandgaps make it more attractive for various applications.[19,20]

TiO$_2$ was reported earlier in nanostructures like nanoparticles, nanotubes, nanorods, nanoflakes, etc. These TiO$_2$ nanotubes (TNTs) are intensively studied for the application of SCs because of their sizeable active area and porous structure.[21]

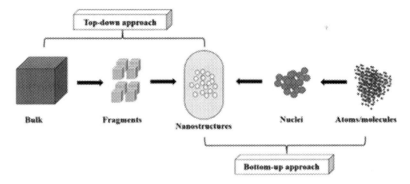

Fig. 1. Approaches for synthesis nanostructures.[21]

TNTs are synthesized using both a top-down approach and a bottom-up approach, as shown in Fig. 1. Anodization is a top-down approach, and hydrothermal is a bottom-up approach commonly used to synthesize TNTs. Among them, anodization is a simple and cost-effective method of controlling the morphology and is also easy compared to the tedious hydrothermal method.[21]

Various reports were published recently on the anodization technique for TNTs fabrication. Recently, Raj *et al.* reported various possible anodization methods for fabricating TNTs by manipulating parameters like anodization voltage, time, and electrolyte concentrations and studied their morphological properties like pore size, diameter, and wall thickness for the applications of electrochemical energy storage.[22] Zohu *et al.* and his team fabricated TNTs array using the anodization technique, and they tuned it by doing a simple cathodic reduction for the fabricated TNTs array electrochemically to improve its electrochemical activity.[23]

Recently, modeling in materials has been trending.[24] It allows researchers to design various complex problems such as nanostructured electrode configurations from simple to complex, and iterate them to find an effective model. Further, implement the same experimentally, making a huge cost-cutting in the material synthesis trial and error phase of material synthesis.[25-27] An effective approach for optimizing and analyzing various physics-based electrochemical models is used to investigate the performance of electrode materials using cyclic voltammetry (CV).

In this work, we have synthesized TNTs using an anodization technique and investigated their physical, structural, and electrochemical characteristics with and without electrochemical etching. The same is modeled by the physics-based finite element method software COMSOL Multiphysics.

2. Numerical Analysis

The simulation structure consists of a single electrolyte domain with the bottom surface representing the TNT electrode surface, as shown in Fig. 2. Here, H, W, L are the height, width, and length of the domain, respectively, h is the height of the TNTs, D_{in} is the inner diameter of the TNTs and D_{out} is the outer diameter of the TNTs.

2.1. *Governing equations*

To perform numerical electrochemical analysis for the TNT electrode, the mass conservation equation for the electrolyte domain was solved for the species transport, expressed as[28]

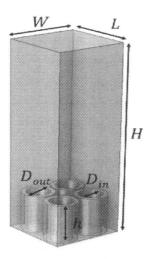

Fig. 2. Simulation model for TNT electrode.

$$\frac{\partial c_i}{\partial t} = -\nabla N_i \quad \text{for } i = 1, 2, \ldots, n, \tag{1}$$

where c_i is the concentration of ith species and N_i is the local molar flux of ith species. For a binary value, $n = 2$, and symmetric electrolyte, the molar flux of ionic species can be expressed in terms of diffusive and electromigration flux of species, given as

$$N_i = -D_i \nabla c_i - z_i u_{m,i} F c_i \nabla \varphi_l. \tag{2}$$

Here, D_i is the diffusion coefficient of ith species, z_i is the charge number of i th species, $u_{m,i}$ is the ionic mobility coefficient of ith species given by the Nernst–Einstein relation ($u_{m,i} = D_i/RT$), F is the Faraday's constant, φ_l is the electric potential of the electrolyte, R is the general gas constant, and T is the temperature.

Additionally, the current conservation equation for the electrolyte was solved and is expressed as

$$\nabla \cdot i_l = 0, \tag{3}$$

where i_l is the current density within the electrolyte given as

$$i_l = -\sigma_l \nabla \varphi_l. \tag{4}$$

Here, σ_l is the electrical conductivity of the electrolyte. At the bottom TNT electrode surface, the current is transferred into the electrolyte purely through the electric double layer, expressed as[29]

$$i_l = C_{dl}\left(\frac{\partial(\varphi_{ext} - \varphi_l - \Delta\varphi_{s,\text{film}})}{\partial t}\right), \tag{5}$$

where C_{dl} is the specific capacitance of the electrode, φ_{ext} is the external electric potential which comes from the applied potential and $\Delta\varphi_{s,\text{film}}$. The film potential is applied to account for the potential drop at the solid–liquid interface. Film potential is given as

$$\Delta\varphi_{s,\text{film}} = R_{\text{film}} i_{dl}. \tag{6}$$

Here, R_{film} is the surface resistance of the TNT electrode surface.

180   S. Harish et al.

2.2. Boundary and initial conditions

Various boundary and initial conditions were considered to solve the governing equations mentioned in the previous sections. At the domain's topmost surface, a constant species concentration boundary condition was applied along with ground electric potential. The symmetry was considered for the four lateral surfaces of the domain. Initially, the whole domain was considered to have uniform species concentration and at zero electric potential.

2.3. Constitutive relationships

Due to the computational limitations in this simulation model, only four TNTs are taken, which is of 150 nm^2 area instead of a total experimental electrode surface of 1 cm^2. Therefore, the total number of nanotubes in a 1 cm^2 area is calculated manually, and it turned out to be 1.78×10^{10}. The various parameters and constraints used in this model are tabulated below with their respective notations in Table 1.

Table 1.   Parameters used in the simulation model.

Parameters	Value	Parameter type
Outer diameter of nanotube (D_{out})	75 nm	Geometry
Inner diameter of nanotube (D_{in})	55 nm	
Height of the nanotube (h)	3000 nm	
Depth of the electrolyte (W)	170 nm	
Length of electrolyte (L)	170 nm	
Height of electrolyte (H)	5000 nm	
Electrolyte diffusion co-efficient of species 1 (D_{c1})	1×10^{-11} m^2/s	Electrolyte
Electrolyte diffusion co-efficient of species 2 (D_{c2})	1×10^{-11} m^2/s	
Electrolyte conductivity (σ_l)	1 S/m	

Table 1. (*Continued*)

Parameters	Value	Parameter type
Starting potential (E_vertex1)	0 V	Electrode surface
End potential (E_vertex2)	0.5 V	
Film resistance (R_{film})	0.25 Ohm·m^2	
Reference film thickness (So)	5 nm	
Film conductivity (σ_{film})	0.001 S/m	
Species 1 concentration ($C_{o,c1}$)	1 mol/m^3	Concentration
Species 2 concentration ($C_{o,c2}$)	1 mol/m^3	
Boundary electrolyte potential ($\varphi_{1,bnd}$)	0 V	Electrolyte potential

2.4. Results and discussion

The simulation of TNTs is carried out using the parameters mentioned in Table 1. As mentioned earlier, it is simulated just for 150 nm^2 area instead of 1 cm^2 for feasible simulation. CV is a powerful electrochemical technique used to estimate the electrochemical performance of the material. The CV plots are depicted in Fig. 3(a) at various scan rates ranging from 10 mV/s to 50 mV/s with a constant potential window of 0–0.5 V applied during simulation. It shows that as the scan rate increases, the corresponding area of the CV plot also proportionally increases. The areal capacitance can be calculated using Eq. (7) where m, *Area*, and $v_2 - v_1$ are the active mass loaded, active substrate area, and potential window applied, respectively, and plotted in Fig. 3(b).

$$C_s = \frac{Area}{m * \text{scanrate} * (v_2 - v_1)}. \quad (7)$$

The CV plot's shape confirms that the capacitance is an EDLC type, and their respective oxidation and reduction peak currents and areal capacitance are tabulated in Table 2.

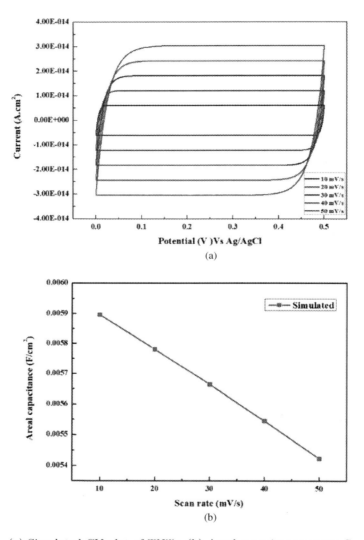

Fig. 3. (a) Simulated CV plot of TNTs. (b) Areal capacitance versus Scan rate plot of Simulated TNTs.

3. Synthesis of TiO$_2$ Nanotubes

3.1. Chemicals and materials

Ethylene glycol (C$_2$H$_6$O$_2$), Ethanol (C$_2$H$_6$O), and Acetone (C$_3$H$_6$O) were purchased from Merk, and Ammonium fluoride (NH$_4$F ≥ 99.99%),

Table 2. Summary of the simulated results.

Scan rate (mV · s^{-1})	Oxidation peak current (fA)	Reduction peak current (fA)	Areal capacitance (pF · cm^{-2})
10	6.1892	−6.0320	1.19393
20	12.2459	−12.1209	1.17662
30	18.2532	−18.6127	1.14710
40	24.3190	−24.4762	1.12289
50	30.6093	−30.3173	1.09816

Titanium foil (Ti 99.7%), Sodium sulphate (Na$_2$SO$_4$) from Sigma-Aldrich. Distilled water is used for solution preparation. All the chemicals were used for synthesis without further purification.

3.2. *Fabrication of TiO$_2$ nanotubes*

The TNTs were fabricated by using the good old technique of anodization.[15] First, mirror polishing the titanium substrate using a machine with different grain-size emery sheets is done. Cleaning the polished substrate is done to remove the contact impurities and debris on the substrate using soap solution, DI water, acetone, ethanol, and DI water, each followed by an ultrasonication bath for 5 min. The cleaned substrates were dried using a portable hair dryer. The anodization parameters for the fabrication of TNTs are mentioned in Table 3.

The electrolyte used for the anodization technique is a mixture of NH$_4$F 4 v/v% and Ethylene glycol (EG) of 96 v/v%, and the setup is as in Fig. 4(a). The mechanism of forming TNTs on a Ti substrate is as depicted in the schematic Fig. 4(b). First, the Ti sheet forms a barrier layer of TiO$_2$ on its surface, and then a pit forms on the surface. Further, the pit grows simultaneously with the barrier layer, forming a pore, later into a tube, with a constant potential applied throughout the experiment. Further, the substrate is washed under running DI water, followed by ultrasonication with ethanol for 5 min to remove any loose debris of the TNTs, and then dried using a portable hair dryer. Finally, the dried sample is

Table 3. Anodization parameters for TNTs.

Target material	Titanium
Technique	Anodization
Substrate	Titanium foil (20 mm × 10 mm)
Substrate thickness	2 mm
Counter electrode	Platinum mesh (20 mm × 20 mm)
Distance between electrodes	25 mm
Operating Voltage (DC)	30 V
Applied current	100 mA
Operating temperature	Room temperature
Time	3 h

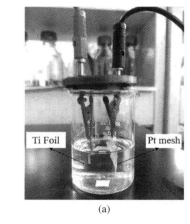

(a)

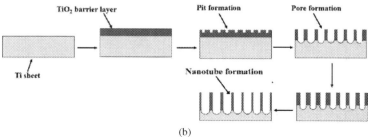

(b)

Fig. 4. (a) Anodization setup for fabricating TNTs, (b) Mechanism of forming TNTs by anodization.

annealed in an air atmosphere at 350°C for 3 h for further characterization and analysis.

For electrochemical etching, a 1 M Na_2SO_4 electrolyte is used with substrate as cathode and platinum electrode as anode for 10 min at an applied voltage of + 3V DC.

3.3. *Characterization*

Field emission scanning electron microscope (FE-SEM) (Thermo Fisher FEI QUANTA 250 FEG) with an operating range of 5–30 kV. Powder X-ray diffractometer (XRD) (German make Bruker D8-Advance) at 2.2 kW with Lynx eye detector. Atomic force microscope (AFM) (Nanosurf easy scan2 23-06-154). Raman spectrometer (Horiba Scientific France XploRATM plus). The electrochemical measurements were done by using a potentiostat (BioLogic SP-200).

4. Results and Discussion

The surface morphology of the TNTs is observed by a FESEM; TNTs are grown perpendicular to the surface of the substrate, and the cross-sectional image is taken by making a cut on the surface of the substrate using a surgical blade. The top and cross-sectional views of the TNTs are shown in Figs. 5(a)–5(d), respectively. The geometry of the grown TNTs is found using the software ImageJ, and the length, outer diameter, and inner diameter of TNTs are in the range of 2900 ± 100 nm, 70 ± 5 nm, and 50 ± 5 nm. The simulation was done by using this geometrical data of TNTs.

The structural properties of fabricated TNTs are studied using a powder X-ray diffractometer (XRD) with a scanning range of 2θ from 20° to 80°, as shown in Fig. 6(a). The metal Ti is oxidized to TiO_2 nanotube film using the anodization technique. The as-prepared sample is characterized as TiO_2, which is confirmed with the JCPDS file number: #00-001-0562 with a tetragonal crystal system with a 141/AMD space group and shows (h k l) peaks of (1 0 1), (1 0 3), (2 0 0), (1 0 5), (2 1 3), (1 1 6), and (1 0 7).[30]

186 S. Harish et al.

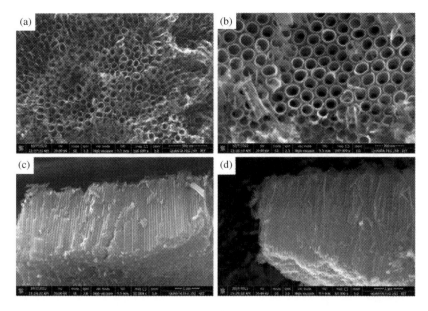

Fig. 5. FE-SEM images of TNTs (a, b) Top view and (c, d) Cross-sectional view.

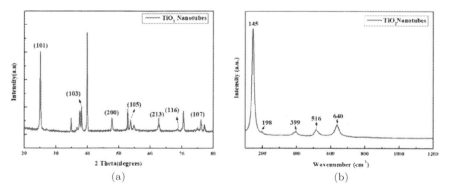

Fig. 6. (a) XRD pattern of as-prepared TNTs (b) Raman spectra of as-prepared TNTs.

TiO_2 shows three different crystalline phases, viz. anatase, rutile, and brookite. The crystalline phase of prepared TNTs is found using a Raman spectrometer with a frequency range of 10–1200 cm^{-1}, as shown in Fig. 6(b). It is observed that the characteristic peaks of

145 cm^{-1}, 198 cm^{-1}, 399 cm^{-1}, 516 cm^{-1}, and 640 cm^{-1} are evident peaks of the anatase crystalline phase.[31]

AFM 2D and 3D imaging is used to find out the parameters like surface roughness of both as-prepared and etched TNTs thin film electrodes with a scanning area of 50 × 50 μm^2 as shown in Fig. 7. Figures 7(a) and 7(c) are 2D and 3D images of as-prepared TNTs thin film and, Figs. 7(b) and 7(d) are etched TNTs' thin film. The tube structure was chipped by etching the TNTs thin film, increasing its roughness, further, making void cavities and improving its porosity and electrode electrolyte interface. Their average surface roughness is 130.23 nm and 168.57 nm for as-prepared and etched TNTs thin film, respectively. The relation between the surface roughness, porosity, and surface area is directly proportional and dependent on each other by maintaining all remaining parameters constant, making a deduced result of the increment of the surface area of etched TNTs thin film by a factor of 1.294 times the as-prepared one.[32,33]

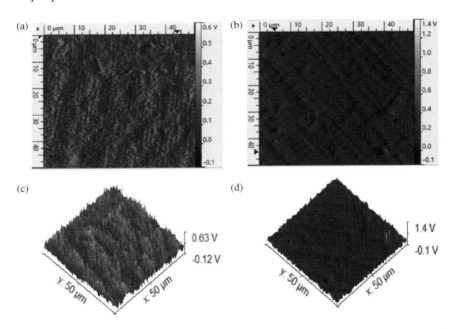

Fig. 7. AFM images of TNTs (a, c) 2D and 3D view of as-prepared TNTs and (b, d) 2D and 3D view of etched TNTs.

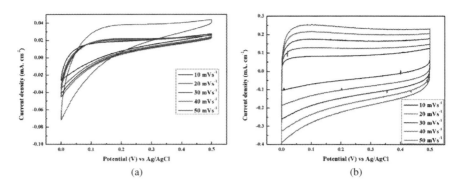

Fig. 8. (a) CV of as-prepared TNTs, (b) CV of electrochemical etched TNTs.

The electrochemical characterization technique of the CV test was used to estimate the prepared TNT's electrochemical performance using various scan rates (mV/s). The potentiostat was used to perform CV tests using the three-electrode system of working electrode (prepared TNTs), counter electrode (platinum rod), and reference electrode (Ag/AgCl) with 1 M Hydrochloric acid (HCL) as electrolyte.

The CV curve is quasi-rectangular, confirming an EDLC-type capacitance behavior.[34] The CV plots of as-prepared TNTs and etched TNTs are shown in Figs. 8(a) and 8(b), respectively. The areal capacitance of the same is depicted in Fig. 9. It offers an areal capacitance of 1.019 F/cm^2 and 12.876 F/cm^2 for as-prepared and etched samples at a standard scan rate of 10 mV/s. It is observed that there is an increase in areal capacitance etched samples compared to as-prepared samples because etching makes a better electrode-to-electrolyte interface, leading to improved capacitance. The areal capacitance values with respective scan rates are tabulated in Table 4.

5. Conclusion

In summary, we fabricated an ordered uniform TiO$_2$ nanotube thin film using a top-down approach to the electrochemical anodization

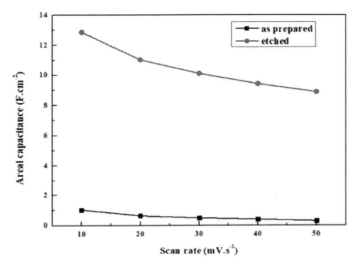

Fig. 9. Areal capacitance versus Scan rate of as-prepared and electrochemical etched TNTs.

Table 4. Areal capacitance obtained at various scan rates.

Scan rate (mV/s)	10	20	30	40	50
Areal capacitance (F/cm^2) as-prepared	1.0193	0.6280	0.4851	0.3968	0.2967
Areal capacitance (F/cm^2) etched	12.8764	11.0311	10.1055	9.4223	8.8926

technique. The consistent nanotubular structure increases the electrode surface area, which is proportional to the capacitance of a supercapacitor. The surface roughness of as-prepared and etched TNTs is 130.23 nm and 168.57 nm, respectively, which improves the surface area of the etched TNTs thin film by a factor of 1.294. The CV reveals that the prepared TiO_2 nanotube thin film shows an EDLC-type capacitance behavior as a quasi-rectangular shape plot. Electrochemical etching improves the interface between the electrolyte and the electrode. The as-prepared TiO_2 nanotubes thin-film electrode shows an areal capacitance of 1.0193 F/cm^2 at 10 mV/s, whereas the etched one shows a 12.8764 F/cm^2 at 10 mV/s using a 1 M HCL. The surface roughness also affects the material's performance as it is proportional to the porosity and surface area. The simulated results were validated using experimental results, and it goes hand-in-hand with the etched TiO$_2$ nanotubes thin-film rather than as-prepared ones.

Acknowledgments

The authors thank Vellore Institute of Technology for providing 'VIT SEED GRANT' No. SG20210068, for carrying out this research work.

ORCID

S. Harish ⊚ https://orcid.org/0000-0002-2114-5159
Muhammad Hamza ⊚ https://orcid.org/0000-0001-5043-1552
P. Uma Sathyakam ⊚ https://orcid.org/0000-0002-6203-0956
A. Senthil Kumar ⊚ https://orcid.org/0000-0001-8800-4038

References

1. J. B. Goodenough, *Ener. Envi. Sci.* **7**, 14 (2014).
2. Z. Zhuang, F. Zhang, Y. Zhou et al., *Mater. Today. Ener.* **30**, 101192 (2022).
3. P. Ma, Z. Zhuang, J. Cao et al., *ACS. Appl. Ener. Mater.* **5**, 6417 (2022).
4. Z. Zhuang, C. Liu, Y. Yan et al., *J. Mater. Chem. A* **9**, 27095 (2021).
5. Z. Zhuang, Y. Tang, B. Ju et al., *Sus. Energy Fuels* **5**, 2433 (2021).
6. A. Yin, L. Yang, Z. Zhuang et al., *Energ. Storage.* **2**, e132 (2020).
7. Poonam, K. Sharma, A. Arora et al., *J. Energy Storage* **21**, 801 (2019).
8. H. Yuan, L. Kong, T. Li et al., *Chin. Chem. Lett.* **28**, 2180 (2017).
9. N. Zhang, Y. Liu, X. Zhang et al., *Funct. Mater. Lett.* **16**, 2351004 (2023).
10. E. Arkhipova, A. Ivanov, S. Savilov et al., *Funct. Mater. Lett.* **11**, 1840005 (2018).

11. S. Najib and E. Erdem, *Nanoscale Adv.* **1**, 2817 (2019).
12. K. Prajapat, M. Dhonde, K. Sahu et al., *J. Photochem. Photobiol C* **55**, 100586 (2023).
13. J. H. Zhu, L. P. Mei, A. J. Wang et al., *Biosens. Bioelectron.* **226**, 115141 (2023).
14. C. Wang, X. Peng, W. Fang et al., *Appl. Surf. Sci.* **614**, 156262 (2023).
15. C. Clement Raj, V. Srimurugan, R. Sundheep et al., *Inorg. Chem. Commun.* **148**, 110286 (2023).
16. M. Pourmadadi, M. Mahdi, E. Rahmani et al., *J. Drug Deliv. Sci. Technol.* **77**, 103928 (2022).
17. A. Farazin, C. Zhang, A. Gheisizadeh et al., *Biomed. Eng. Adv.* **5**, 100075 (2023).
18. C. C. Raj, V. Srimurugan, A. Flamina et al., *Mater. Chem. Phys.* **248**, 122925 (2020).
19. B. S. Lee, *Polymers* **12**, 2035 (2020).
20. D. R. Eddy, M. D. Permana, L. K. Sakti et al., *Nanomaterials* **13**, 704 (2023).
21. S. Harish and P. U. Sathyakam, *J. Energy Storage* **52**, 104966 (2022).
22. C. C. Raj and L. Neelakantan, *J. Electrochem. Soc.* **165**, E521 (2018).
23. H. Zhou and Y. Zhang, *J. Power. Sources* **239**, 128 (2013).
24. C. Hao, X. Wang, Y. Yin et al., *J. Electron. Mater.* **45**, 515 (2016).
25. P. Chinnasa, W. Ponhan and W. Choawunklang, *J. Phys. Conf. Ser.* **1380**, 012101 (2019).
26. S. Aderyani, P. Flouda, S. A. Shah et al., *Electrochem. Acta.* **390**, 138822 (2021).
27. S. Aderyani, S. A. Shah, A. Masoudi et al., *ACS Nano* **14**, 5314 (2020).
28. M. S. Kilic, M. Z. Bazant and A. Ajdari, *Soft. Matter Phys.* **75**, 01 (2007).
29. C. Lin, J. A. Ritter, B. N. Popov et al., *J. Electrochem. Soc.* **146**, 3168 (1999).
30. H. Attouche, S. Rahmane, S. Hettal et al., *Optik* **203**, 01 (2020).
31. J. Wang and Z. Lin, *Chem. Mater.* **20**, 1257 (2008).
32. T. Sakai and A. Nakamura, *Ear. Plan. Spa.* **57**, 71–76 (2005).
33. M. Mohammadi, S. Shadizadeh, A. Manshad et al., *J. Petro. Explo. Prod. Tech.* **10**, 1817–1834 (2020).
34. D. Majumdar, *J. Electroanal. Chem.* **880**, 114825 (2021).

Chapter 11

Ton-scale preparation of single-crystal Ni-rich ternary cathode materials for high-performance lithium ion batteries

Yongfu Cui [*,‡], Jianzong Man [†], Leichao Meng [*], Wenjun Wang [*], Xueping Fan [*], Jing Yuan [*] and Jianhong Peng [*]

*Qinghai Provincial Key Laboratory of Nanomaterials and Nanotechnology Qinghai Minzu University, Xining 810007, P. R. China
†Shandong Provincial Key Laboratory of Chemical Energy Storage and Novel Cell Technology, and School of Chemistry and Chemical Engineering Liaocheng University, Liaocheng 252000, P. R. China
‡yongfucui@163.com

Single-crystal nickel (Ni)-rich cathode materials (Li[Ni$_x$Co$_y$Mn$_{1-x-y}$]O$_2$, (NCMs)) of lithium ion batteries (LIBs) have displayed promising application potential due to the merits of stable structure, minor side reaction, and high energy density. The Ni$_x$Co$_y$Mn$_{1-x-y}$(OH)$_2$ as the precursor of single-crystal NCM faces the issues of tedious preparation process and serious pollutant emission for traditional preparation methods during the industrial preparation. Herein, an improved continuous two-step spray pyrolysis strategy is adopted to prepare the precursor of single-crystal NCM. Combining with the industrial devices, ton-scale preparation of single-crystal NCM is realized after dynamic lithiation post-treatment. This improved strategy not only shortens the preparation process, but also reduces metal segregation and sintering temperature, effectively balancing the cost control and production efficiency. The samples sintering at 850°C show uniform morphology with a diameter of 4.5 μm and delivers an initial discharge capacity of 169 mAh/g at 0.1 C after 100 cycles. This work provides a new route for the industrial preparation single-crystal NCM cathode materials of LIBs.

---

‡Corresponding author.
To cite this article, please refer to its earlier version published in the Functional Materials Letters, Volume 16(8), 2340035 (2023), DOI: 10.1142/S1793604723400350.

Keywords: Single-crystal ternary cathode materials; Li[Ni$_{0.6}$Co$_{0.2}$Mn$_{0.2}$]O$_2$; spray pyrolysis; ton-scale production.

1. Introduction

Although lithium ion batteries (LIBs) with mature technology have been applied in many aspects in life such as electric vehicle and portable electrical devices, higher energy density and longer lifetime have been pursued all the time.[1–3] The cathode materials are crucial to the safety and energy density of LIBs.[4,5] Single-crystal nickel (Ni)-rich cathode materials (Li[Ni$_x$Co$_y$Mn$_{1-x-y}$]O$_2$, (NCMs)) have drawn much more attention in recent years due to the merits of high energy density, stable structure, minor side reaction, and low cost.[6–8] Compared with polycrystalline Ni-rich NCM, single-crystal NCM possesses excellent structural strength against microcracking. However, the scaled preparation of single-crystal NCM in the industrial field has hindered its development further.[9–12] Different from the experiments in the laboratory, the issues of cost control, pollutant emission and production efficiency are also key factors except for the favorable electrochemical performance during the preparation of single-crystal NCM in the industrial field. Traditional methods including the co-precipitation method,[13] molten-salt method,[14] the versatile hydrothermal methods,[15] and spray pyrolysis method[16] have been selected to prepare the single-crystal NCM, and the as-prepared single-crystal NCM shows excellent electrochemical performance. For instance, Qian *et al.* synthesized single-crystal LiNi$_{0.83}$Co$_{0.12}$Mn$_{0.05}$O$_2$ cathode material with an excellent capacity of 171.6 mAh/g at 0.3 C over 500 cycles, using co-precipitation and high-temperature solid-phase methods.[13] Guan *et al.* synthesized single-crystal NCM using an industrially applicable molten-salt approach, which showed a high specific capacity of 183 mAh/g at 0.1 C and excellent capacity retention of 172 mAh/g at 1 C over 300 cycles.[14] Lei *et al.* synthesized single-crystal NCM with diameters of ~800 nm and a high capacity of 183.7 mAh/g at 0.2 C using a versatile hydrothermal method.[15] Nevertheless, there are some challenges to preparing the single-crystal NCM, particularly Ni-rich ternary cathode materials, which is attributed to the uncontrollable homogeneity of three different elements (Ni, Co, and Mn) in the single-crystal NCM during the synthetic process. Fan et al. monitored a lot of cracks in the spherical secondary agglomerates Ni-rich Li[Ni$_x$Co$_y$Mn$_{1-x-y}$]O$_2$ (0.6 ≤ x ≤ 0.95) cathode materials, while most layered structures were still retained in single-crystal NCM under the same conditions.[17] However, several issues including bad electrode/electrolyte interface stability,[18] huge volume change,[19,20] unstable structure,[21] poor heat resistance,[22] and inferior cyclic performance[14] for the preparation of Ni-rich Li[Ni$_x$Co$_y$Mn$_{1-x-y}$]O$_2$ (0.6 ≤ x ≤ 0.95) cathode materials via traditional methods have not been addressed well. Above all, although the single-crystal NCMs have been reported with excellent electrochemical performance, balancing the factors that electrochemical performance, production efficiency and pollutant emission should not be overlooked.

In this work, a new pilot plant with a capacity of 30 tons per year was designed to prepare the hybrid oxides NiO-MnCo$_2$O$_4$-Ni$_6$MnO$_8$ (HONCM) precursors through metal chlorate pyrolytic reaction. As shown in Fig. 1, the continuous two-step spray pyrolysis

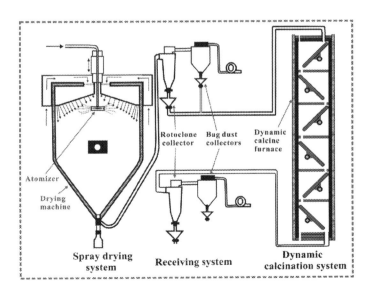

Fig. 1. Schematic illustration of spray drying and pyrolysis process.

has been designed and then used for the preparation of HONCM precursors. The equipment consists mainly of spray drying system, dynamic calcination system, and receiving system. The realistic photos are displayed in Fig. S1, indicating the facticity of the continuous two-step spray pyrolysis system. Firstly, the precursor metal solutions are atomized into aerosol droplets by pressure atomization, and then the droplets are directly sprayed into the drying equipment, which undergoes a series of physical processes including water vaporization, dry deposition, and drying. As the low-temperature spray drying process progressed, deionized water was evaporated from the atomized droplets, and then the chloride salts gradually lost crystal water. Remarkably, it is favorable to prepare the chloride salts particle with little size distribution and good dispersion at a lower drying temperature and in a reduced drying time. The spray drying particles were collected by rotoclone collectors, and the smaller micro-sized powders were also recycled by the bag dust collectors at the same time. The obtained particles after drying were dynamically calcinated in the calcination system through a feeding machine subsequently. The chambers of the dynamic calcination system are designed in the shape of a slide, which is beneficial to extend the residence time of the particles to about thirty seconds in the spray drying process. Finally, the particles with favorable flowability and uniform particle size are collected by the cyclone collector. All the waste pyrolysis gas and solvent vapor generated in the preparation process were absorbed by alkaline solution, realizing non-pollution production. Furthermore, continuous two-step spray pyrolysis production cost is reduced by 40% compared to the commercial co-precipitation method after repeated calculations, providing feasibility for large-scale production of cathode materials. The single-crystal NCM prepared by the continuous two-step spray pyrolysis system and novel technology delivers a reversible capacity of electrochemical properties with a stable discharge capacity and an excellent rate capability.

2. Experimental Methods

2.1. *Raw materials and medicines*

The regents of $NiCl_2·6H_2O$, $MnCl_2·4H_2O$, and $CoCl_2·6H_2O$ were industrial-grade and purchased from the market. All the chloride salts are industrial-grade. All chemicals used in the experimental procedures of this work were purchased from Tianjin Kermel Co., Ltd, which were of analytical grade and used directly without any purification. Distilled water was employed in all synthesis and treatment processes.

2.2. *Synthesis of precursor HONCM*

Firstly, the $NiCl_2$, $MnCl_2$, and $CoCl_2$ were dissolved in the deionized water with the molar ratio of 6:2:2. The solution was stirred for 1 h to form an even solution and kept at 150 g/L. The precursor HONCM was prepared with a spray pyrolysis equipment. The spray pyrolysis was fed with the flow rate of 30 L/min. Subsequently, the drying powders were distinguished by cyclone separator. The powders with different sizes were transported into the pyrolysis kiln. The physical and chemical components are shown in Table S1.

2.3. *Preparation of single-crystal NCM*

The single-crystal NCM was prepared by a typical lithiation process. The precursors HONCM and LiOH were mixed with a molar ratio of 1:1.05. The mixture suffered calcination at 450°C for 2 h initially, then it was heated at 650°C for 6 h in the second stage. As the temperature was raised to 870°C, the mixture was heated for 10 h in the last stage. The temperature rate is kept at 1°C/min. after the furnace was cooled to indoor temperature, the obtained sample was single-crystal NCM.

2.4. Material analysis

The crystal structure of precursor HONCM and single-crystal NCM samples were analyzed by X-ray diffraction (XRD, Rigaku D/max-Ultmer, Rigaku Corporation, Japan) using a Cu Kα radiation over a scattering angle from 10° to 80°. The size of single-crystal NCM particle distribution was characterized by a laser particle-size analyzer (LPSA, Zetasirer nano ZS, MALVERN, UK). The element distribution of single-crystal NCM samples was carried out via an energy-dispersive X-ray spectrometer (EDS) equipped attached to the SEM. The morphology of single-crystal NCM samples was observed via a field-emission scanning electron microscope (FE-SEM, SUPRA 55 SAPPHIRE, ZEISS).

2.5. Electrochemical measurements

The 2025 type coin-cell was assembled in the glove box with the content of water and oxygen bellowing 0.1 ppm. Single-crystal NCM, carbon black, and PVDF (mass ratio by 8:1:1) and NMP were mixed to form a slurry and then coated on the aluminum foil. After drying in vacuum for 12 h, the electrode was prepared by cutting the Al foil into a disk with a diameter of 14 mm. A solution of 1.0 M LiPF6 in ethylene carbonate: dimethyl carbonate (EC : DMC by 1:1 in volume) was applied as an electrolyte. LAND-CT2001A battery test system was used to carry out a galvanostatic charge–discharge test. The test of cyclic voltammetry (CV) was carried out by the CHI760E electrochemical workstation between 2.8 V and 4.3 V.

3. Results and Discussion

As shown in Fig. 2, the crystal structure of precursor HONCM and obtained single-crystal NCM is characterized by XRD. The samples of precursor HONCM sintering at 750°C and 780°C have the same diffraction peaks, suggesting the unchanged crystal structure under a relatively low temperature. When the sintering temperatures are increased to 800°C, 850°C, and 900°C, the diffraction peaks of NiO

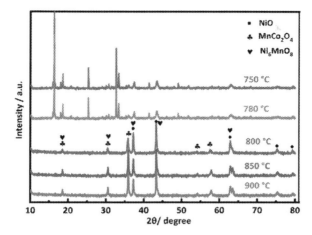

Fig. 2. XRD patterns of precursor HONCM calcinated at different temperature.

Fig. 3. SEM images of precursor $Ni_xCo_yMn_{1-x-y}(OH)_2$ at different calcination temperatures (a) 750°C; (b) 780°C; (c) 800°C; (d) 850°C; (e) 870°C; (f) 900°C.

(JCPDS # 44-1159), $MnCo_2O_4$ (JCPDS #84-0482) and Ni_6MnO_8 (JCPDS #49-1295) are observed. Besides, there are no other peaks in the XRD patterns, demonstrating the unicity of samples after sintering.

Figure 3 shows the micromorphology and crystal structure of precursor HONCM at different temperatures from 750°C to 900°C.

200   Y. Cui et al.

As shown in Figs. 3(a)–3(f), the particle size of precursor HONCM increases with the increase of lithiation temperature. The D50 of precursor HONCM sintering at 850°C is about 4.5 μm. The size of precursor HONCM particles is suitable for further lithiation process. The single-crystal NCM was prepared via calcination using the precursor HONCM (850°C) and LiOH as the precursor and Li source, respectively. LiOH is the lithium source to synthesize Ni-rich cathode materials due to their relatively lower melting point compared to Li_2CO_3, which is in favor of reducing the calcination temperature of Ni-rich cathode materials. Figure 4(a) shows the SEM images of single-crystal NCM microspheres calcinating at 850°C. The particle size distribution of the single-crystal NCM sample is narrow and uniform, which proves that there is a good pyrolysis process. The element distribution of single-crystal NCM is shown in Fig. 4(c), Ni, Co, Mn, and O are uniformly dispersed in

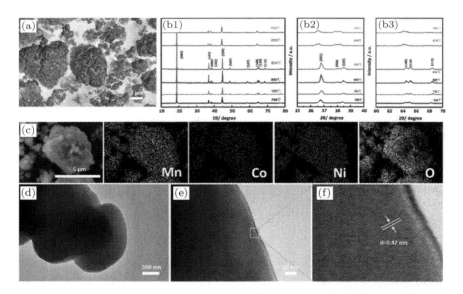

Fig. 4. (a) SEM morphologies of single-crystal NCM; (b) XRD patterns of single-crystal NCM particles by different calcination temperatures and the partially enlarged patterns; (c) elements distribution of S850; (d) TEM image of single-crystal NCM; (e) and (f) high-resolution TEM images of single-crystal NCM.

single-crystal NCM, verifying the successful preparation of single-crystal NCM. The physical parameters are within the references ranges, indicating the successful preparation of single-crystal NCM (Table S2).

The phase structure of single-crystal NCM calcinating at different temperatures is measured by XRD. As shown in Figs. 4(b-1), the single-crystal NCM exhibits a similar α-NaFeO$_2$[24] lamellar crystal structure. The typical characteristic diffraction peaks of single-crystal NCM samples could be assigned to (003), (101), (006), (102), (104), (105), (107), (108), (110), and (113) planes of α-NaFeO$_2$ (JPCDS No. 09-0063).[23] It can be found that the diffraction peaks in the XRD patterns of samples at 750°C, 780°C, 800°C, and 850°C are completely consistent with them, and the diffraction peaks (101) and (113) of samples at 870°C and 900°C are weak in Figs. 4(b-2) and 4(b-3), which may be due to the phase structure of single-crystal NCM changing for the excessive calcination temperature. No other impurity peaks were observed in the patterns, which means that the precursor HONCM is lithiated well. More remarkably, the distinct splitting of the (006)/(102) and (108)/(110) doublets demonstrates a highly-ordered layered configuration of single-crystal NCM,[24,25] the lattice parameters of the samples heated at different temperatures also show similar results (Table S3), especially at 850°C. In addition, this typical single-crystal NCM sample was named S850. From Fig. 3(d), the aggregation of single-crystal NCM is observed and the particle size of single-crystal NCM is about 2 μm, which is attributed to the size decrease in the high temperature atmosphere. The high-resolution TEM reveals uniform interplanar distances (0.47 nm), corresponding to the (003) crystal planes of the layered structure.

The lithiation/delithiation processes of the sample S850 were investigated by CV curves between 2.8 V and 4.3 V at a scan rate of 0.1 mV/s for the initial three cycles, as exhibited in Fig. 5(a). It can be observed that the first redox peaks obviously differ from the subsequent 2nd and 3rd in the CV curves. It is distinctly presented in Fig. 5(a) that the major oxidation peak transforms from 3.78 V to 3.88 V in the first cycle, which is attributed

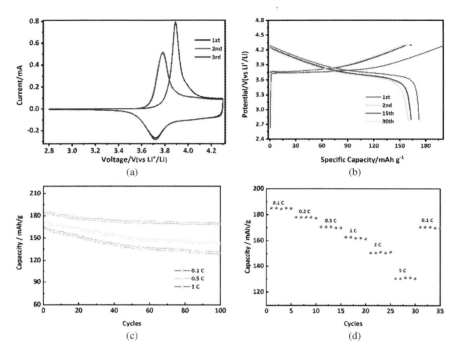

Fig. 5. (a) Cyclic voltammograms for the first three cycles, (b) charge–discharge profiles at 0.1 C for the 1st, 2nd, 15th, and 30th cycle of sample S850 electrode, (c) cycling performance at 0.1 C, 0.5 C, 1 C, and (d) rate capability from 0.1 C to 5 C in the voltage of 2.8–4.3 V of sample S850 electrode.

to the irreversible reaction between the electrolyte and electrode, and the curves are Overlapped for the second and third cycles. Only a pair of redox peaks are displayed in each CV curve, suggesting that no multiphase reactions existed during electrochemical cycles. As displayed in Fig. 5(b), the initial charge capacity of the sample S850 electrode reaches up to 204.5 mAh/g at a rate of 0.1 C, and the electrode has a higher discharge capacity of 184 mAh/g accompanied by a higher initial Coulombic efficiency (CE) (90%). Besides, the charge–discharge profiles of the 15th and 30th cycles are nearly overlapped, indicating that the sample S850 electrode has excellent cycling stability. The long-term cycling performance of S850 was tested at 0.1 C, 0.5 C, and 1 C, respectively, (Fig. 5(c)). The single-crystal NCM also delivers a capacity retention of 163.5 mAh/g after 400 cycles, showing a capacity loss ratio of 0.012% per cycle. The initial discharge capacity is decreased with the increase of applied current density. The S850 delivers capacities of 185, 170, and 163 mAh/g at 0.1, 0.5, and 1 C, respectively. After 100 cycles, the capacities are obtained at 169, 143, and 130 mAh/g, showing a favorable cycling stability. Figure 5(d) shows the rate performance of S850 from 0.1 C to 5 C. The S850 delivers the capacity of 130 mAh g^{-1} at the current density of 5 C, suggesting the outstanding dynamic at high current density.

4. Conclusions

For the preparation of Ni-rich single-crystal NCM on a large scale, a modified strategy of improved continuous two-step spray pyrolysis method was used. Combining with the industrial equipment, ton-scale preparation of single-crystal NCM is realized with low cost and reduced pollutant emission. The precursor HONCM samples sintering at 850°C show uniform particle size and larger particle size distribution. The sample S850 electrode shows uniform well-crystallized fine particles and delivers excellent electrochemical performance. The S850 delivers a capacity of 130 mAh g^{-1} at 5 C. This strategy provides an applicable strategy for the practical preparation of single-crystal NCM on a large scale.

Acknowledgments

This work was supported by the Natural Science Foundation of Qinghai Province (No. 2020-zj-910).

ORCID

Yongfu Cui ◉ https://orcid.org/0000-0003-4785-3695
Jianzong Man ◉ https://orcid.org/0000-0002-3963-5156
Leichao Meng ◉ https://orcid.org/0000-0003-1038-8785
Wenjun Wang ◉ https://orcid.org/0000-0001-6216-421X

Xueping Fan ⊚ https://orcid.org/0009-0006-7344-435X
Jing Yuan ⊚ https://orcid.org/0000-0001-9325-4891
Jianhong Peng ⊚ https://orcid.org/0000-0002-9636-8162

References

1. H. D. Yoo *et al.*, *Mater. Today* **17**(3), 110 (2014).
2. Z. P. Cano *et al.*, *Nat. Energy* **3**, 279 (2018).
3. M. Winter, B. Barnett and K. Xu, *Chem. Rev.* **118**, 11433 (2018).
4. S.-T. Myung *et al.*, *ACS Energy Lett.* **2**, 196 (2017).
5. G.-H. An, H. Kim and H.-J. Ahn, *Appl. Surf. Sci.* **463**, 18 (2019).
6. X. Ma *et al.*, *Chem-US* **8**(7), 1944 (2022).
7. P. Yue *et al.*, *Powder Technol.* **237**, 623 (2013).
8. L. Jin *et al.*, *Chem. Soc. Rev.* **48**, 3015 (2019).
9. T. Chen *et al.*, *J. Power Sources* **374**, 1 (2018).
10. X.-M. Fan, *Adv. Funct. Mater.* **32**(6), 2109421 (2021).
11. W. H. Kan *et al.*, *Chem-US* **4**, 2108 (2018).
12. L. Feng *et al.*, *J. Mater. Chem. A* **6**, 12344 (2018).
13. Q. Guo *et al.*, *New J. Chem.* **45**, 3652 (2021).
14. G. Qian *et al.*, *Energy Storage Mater.* **27**, 140 (2020).
15. L. Wang *et al.*, *J. Alloys Compd.* **674**, 360 (2016).
16. J. Zhu *et al.*, *J. Power Sources* **464**, 228207 (2020).
17. X. Fan *et al.*, *Nano Energy* **70**, 104450 (2020).
18. M.-H. Kim *et al.*, *J. Power Sources* **159**, 1328 (2006).
19. J. H. Lee *et al.*, *Energy Environ. Sci.* **9**, 2152 (2021).
20. H. Kim *et al.*, *Nano Lett.* **15**, 2111 (2015).
21. Z. Xu *et al.*, *J. Mater. Chem. A* **6**, 21859 (2018).
22. D. Rathore *et al.*, *J. Electrochem. Soc.* **168**, 120514 (2021).
23. S. A. Needham *et al.*, *J. Power Sources* **174**, 828 (2007).
24. K. M. Shaju, G. V. Subba Rao and B. V. R. Chowdari, *Electrochim. Acta* **48**, 145 (2002).
25. J. Li *et al.*, *J. Electrochem. Soc.* **162**(7), A1401 (2015).

Chapter 12

Investigation on the influence of Zn content on the structural, optical, morphological and electrical properties of ternary compound $Cd_{1-x}Zn_xS$ window layer for CdTe solar cell

Ki Tong Hun[*], Kim Hyon Chol[†,‡] and Jo Hye Gang[†]

[*]Faculty of Electronics, Kim Chaek University of Technology
Pyongyang, Democratic People's Republic of Korea
[†]Laboratory of Photovoltaics, Faculty of Electronics
Kim Chaek University of Technology, Pyongyang
Democratic People's Republic of Korea
[‡]khc86217@star-co.net.kp

Ternary compound $Cd_{1-x}Zn_xS$ thin films with different zinc compositions were deposited by the chemical bath deposition method, as a window layer for CdTe solar cells. The influence of Zn content on the structural, optical, morphological and electrical properties of $Cd_{1-x}Zn_xS$ was studied from X-ray diffraction (XRD), SEM and UV-vis spectroscopy. We have considered the transformation of the crystal phase, change in the crystallite size and lattice parameters of $Cd_{1-x}Zn_xS$ thin film with the change of Zn content through XRD studies. SEM measurement revealed that the size of $Cd_{1-x}Zn_xS$ particles and surface roughness of thin film decreases with the increase of Zn composition. It was found that the band gap value obtained from transmittance spectrum became wider with the increase of Zn contents, which indicates the increase of film transmittance. The electrical resistivity of $Cd_{1-x}Zn_xS$ thin films increased as Zn content x increases. Finally, we measured the performances of CdTe solar cells using $Cd_{1-x}Zn_xS$ thin films as a window

---

[‡]Corresponding author.
To cite this article, please refer to its earlier version published in the Functional Materials Letters, Volume 16(8), 2340038 (2023), DOI: 10.1142/S1793604723400386.

layer and confirmed that desirable Zn content was in the range $x = 0.5$–0.6.

Keywords: $Cd_{1-x}Zn_xS$; CdTe solar cell; Zn content; window layer; crystal phase; band gap.

1. Introduction

In recent years, thin film CdTe solar cell has been paid much attention from both the research area and the photovoltaic industry due to its advantages of low cost and high efficiency.[1] In order to improve the efficiency of solar cell, Zn-doped or alloyed $Cd_{1-x}Zn_xS$ materials with the wide band gap have drawn tremendous research interest for the replacement of CdS window material with the relatively small value of E_g (2.42 eV) which absorbed considerably solar energy in the region of shorter wavelength (< 500 nm) to decrease the current density of heterojunction solar cells.[2,3] The adjustment of band gap is very useful to design a window layer in the fabrications of heterojunction solar cells and the optoelectronic devices.[4,5] Zinc and cadmium are suitable to form a solid solution due to their isovalency and very similar electronegativity (Cd: 1.69, Zn: 1.65).[6] The addition of Zn element into CdS improves the short circuit current density and open circuit voltage to increase the efficiency of hetero-junction-based solar cells such as CdTe and CIGS,[7] decreasing the absorption losses of the window materials in p-n junction devices and the lattice mismatch.[4]

Over the past three decades, there have been a number of reports on the deposition and properties of $Cd_{1-x}Zn_xS$ ternary compound with various ranges of Zn concentration (low Zn content,[4,8,10] midst content[11,12] and overall Zn content[1,3,5,6,13,15]), using different methods like chemical bath deposition,[6,7,10,16,17] spray pyrolysis,[3] SILAR,[18] vacuum evaporation,[19,20] MOCVD,[13] RF sputtering,[21] vapor zinc chloride treatment,[22] etc. Besides these, several reports have been published not only on the effect of substrate[16] and heat treatment[12,14,20,21,23] but also on the electronic transport,[24] first-principle approach[25,26] of $Cd_{1-x}Zn_xS$ material. Even though many authors reported their research results on the synthesis and properties of

$Cd_{1-x}Zn_xS$ in the form of particle or film, for the purpose of using as a window layer of solar cell, photo-catalyst[9,27,28] and other optoelectronic devices, the approach of this field still remains to be a challenging task. In this paper, we investigated the influence of Zn content on the structural, optical, morphological and electrical properties of ternary compound $Cd_{1-x}Zn_xS$ window layer for CdTe solar cell.

2. Experimental

$Cd_{1-x}Zn_xS$ thin films were prepared by the CBD method on FTO-coated glass substrates (2 cm × 2 cm). Before deposition, the substrate surfaces were ultrasonically washed several times with cleaning agents such as alcohol, acetone and deionized water, and then dried at the temperature of 80°C in a drying oven.

Aqueous solutions of $CdCl_2$, $ZnCl_2$ and $(NH_2)_2CS$ were used as sources of Cd, Zn and S, respectively. The solution concentrations of thiourea and ammonium chloride were kept as 6 mM and 20 mM, respectively. In all cases, S/(Cd + Zn) molar ratio was fixed as 6, keeping the relation: $[CdCl_2] + [ZnCl_2] = 1$ mM. In order to change the value of $x(= Zn/(Cd + Zn))$ from 0 to 1, the concentration of $ZnCl_2$ was varied from 0 to 1 mM as shown in Table 1.

The chemicals used in the fabrication process of thin films were up to analytical grade. For preparing the solution, FTO glasses were immersed in it. Bath temperature was increased with the rate of 1.5°C/min. When the deposition temperature reached up to 85°C, ammonia was added drop by drop into the solution to adjust pH. The deposition time was 60 min. After the film deposition,

Table 1. Solution concentration of $CdCl_2$ and $ZnCl_2$ to change Zn composition, x.

Composition, x	0.0	0.1	0.2	0.3	0.4	0.5	0.6	0.7	0.8	0.9	1.0
$[CdCl_2]$, mM	1.0	0.9	0.8	0.7	0.6	0.5	0.4	0.3	0.2	0.1	0.0
$[ZnCl_2]$, mM	0.0	0.1	0.2	0.3	0.4	0.5	0.6	0.7	0.8	0.9	1.0

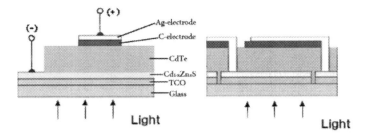

Fig. 1. Superstrate structure of CdTe solar cell.

glasses were withdrawn from the bath solution, and then cleaned ultrasonically and dried in warm and fresh air.

For some of the above prepared $Cd_{1-x}Zn_xS$ films, the complete fabrication of CdTe solar cells was carried out.

Figure 1 shows the superstrate structure of the fabricated solar cell.

CdTe layers were deposited by the close space sublimation (CSS) technique in which the temperatures of source and substrate were 650°C and 550°C, respectively. The chamber was evacuated down to 10^{-3} mbar. The deposition was done for 5 min and then the source and substrate lamps were switched off for cooling down to 100°C before opening the chamber. All the samples were annealed at 400°C for 30 min under atmosphere conditions. These samples underwent $CdCl_2$ heat treatment at 400°C, NP etching and the formation of back contact.[29]

X-ray diffraction (XRD) measurements were performed on D8-ADVANCE diffractometer (scanning range 0°–80°, $\lambda = 1.5406$ □, 40 kV at 40 mA, scanning rate 0.06/s). The optical transmittance of $Cd_{1-x}Zn_xS$ films was measured in the wavelength range of 300–800 nm on a Jasco V-670 UV–Vis–NIR spectrophotometer. Surface morphology of $Cd_{1-x}Zn_xS$ films were observed by SEM (QUANTA 200) apparatus attached to EDX. The solar cell performances were measured under AM1.5 illumination.

3. Results and Discussion

3.1. X-ray diffraction (XRD) studies

3.1.1. Formation process of $Cd_{1-x}Zn_xS$ thin film

The heterogeneous reaction causes the uniform distribution of crystallites in Cd–S lattice by the "ion by ion" process to form the pure hexagonal, cubic phase or admixed hexagonal phase depending on the fabrication condition.[4] During the growth of crystal in solution, the solvation of cluster nuclei affects the crystal phase and growth direction as well as the growth rate.[11] The formation mechanism of $Cd_{1-x}Zn_xS$ thin film is related to the production of coordination complexes with Cd^{2+} and Zn^{2+} coordinated to S.[11]

The overall reaction formula to form $Cd_{1-x}Zn_xS$ film is as follows[5,11]:

$$(1-x)Cd(NH_3)_4^{2+} + xZn(NH_3)_4^{2+} + S^{2-} \rightarrow Cd_{1-x}Zn_xS + 4NH_3\uparrow. \quad (1)$$

3.1.2. Change of the crystal phase

Figure 2 shows the result of XRD analysis of $Cd_{1-x}Zn_xS$ thin films with different Zn contents in the range of $0 < x < 1$. As shown in this figure, $Cd_{1-x}Zn_xS$ samples with the Zn content $0.0 < x < 0.4$ have six peaks corresponding to (100), (002)/(111), (101), (110)/(220), (103)/(200), (112)/(311) crystal plane with the preferential orientation along the (002) plane. These peaks represent the reflections at 2θ angles of 24.8°, 26.8°, 28.2°, 43.7°, 47.7°, 52.1° characteristic of pure hexagonal CdS phase.

The insertion of Zn into Cd–S lattice leads to the mixed structure of the hexagonal and cubic phases. In other words, the pure cubic ZnS phase was not produced at the lower range of Zn content, the homogeneous $Cd_{1-x}Zn_xS$ solid solution is formed by the incorporation of Zn into the hexagonal CdS structure.[6]

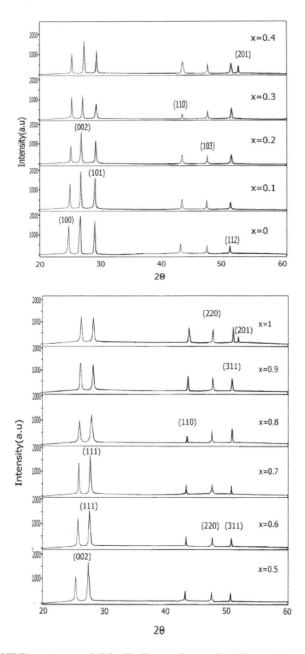

Fig. 2. XRD patterns of $Cd_{1-x}Zn_xS$ samples with different Zn content x.

From $x = 0.5$, the peak intensity of thin film gets stronger due to the overlap of (002) and (101) peaks. The increasing trend of the peak intensity is caused by the substitution of Zn^{2+} instead of Cd^{2+} ions.[4] On the other hand, at $x > 0.6$, the peak intensity decreases as Zn^{2+} ions enter into Cd–S lattice in the form of the substitution and interstitial.

The difference in the atomic radius and crystal structure between CdS and ZnS materials is a main factor which limits the creation of solid solution at $x < 0.4$.[6] The interstitial of Zn into the Cd–S lattice caused the mixture of cubic phase and hexagonal phase, which were observed at Zn content $0.4 < x < 0.5$. The peak with the highest intensity was changed from CdS hexagonal phase (002) to ZnS cubic phase (111) in this range.

3.1.3. Variation of the crystalline size and lattice constant

The intensities and positions of XRD diffraction peak of $Cd_{1-x}Zn_xS$ film depend strongly on the properties of doping material.[4] With the increase of Zn composition from 0.0 to 1.0, the peak intensity decreases and 2θ angles are shifted slightly towards the higher value, representing the decrease of the lattice constant. The continuous shift of XRD pattern represents the formation of $Cd_{1-x}Zn_xS$ solid solution by the implantation of Zn into the CdS lattice.[9] This variation can be attributed to the substitution of Zn^{2+} ions (radius 0.74) with a smaller size than Cd^{2+} ions (radius 0.97), which decreases the crystalline size and lattice constant.[6] Meanwhile, the increase of 2θ angle corresponds to the contraction of inter-planar distance.[16]

The average crystallite size of $Cd_{1-x}Zn_xS$ thin film can be calculated from 2θ angle and full width at half maximum (FWHM) of XRD diffraction peaks using the Scherrer equation as follows[3,4]:

$$D = \frac{0.9\lambda}{\beta \cos\theta}, \qquad (2)$$

where D is the average crystallite size, 0.9 is a value of the Scherer's constant, β is the FWHM in radians and λ is a value of wavelength of X-rays (1.5406 Å).

The change of FWHM value reveals that Zn doping affects the microstructure such as the crystallinity and lattice parameter.

The average crystallite size of $Cd_{1-x}Zn_xS$ thin film decreased from CdS to ZnS, which could be proved by the decrease of peak intensity, broadening of FWHM and the shift of 2θ angle towards the higher value.[6]

The inter-planar distances d were calculated from the Bragg formula ($n\lambda = 2d\sin\theta$), and then the lattice constants (a and c) could be obtained from the following equations.[3,16]

$$\frac{1}{d^2} = \frac{4}{3}\left(\frac{h^2 + hk + k^2}{a^2}\right) + \frac{l^2}{c^2} : \text{for hexagonal phase,} \qquad (3)$$

$$\frac{1}{d^2} = \frac{h^2 + k^2 + l^2}{a^2} : \text{for cubic phase,} \qquad (4)$$

where λ is the X-ray wavelength, θ is the diffraction angle and d indicates the inter-planar distance of indices (h, k, l), where h, k and l are the Miller index; in the case of the hexagonal phases, a and c were calculated from (100) and (002), respectively.

The gradual decrease of lattice parameter 'a' was observed with the increase of Zn content. This trend indicates the change in the composition of $Cd_{1-x}Zn_xS$ solid solution by the higher degrees of substitution of Zn, considering that lattice parameters follow Vegard's low, which also shows that the atom arrangements get compact in the crystals.[9]

The peak broadening observed from the XRD measurements indicates the decrease of crystallite size with the increase of Zn content.[11] The substitution of Zn^{2+} at low-doping concentration disturbs the Cd–S lattice and subsequently arrests the nucleation and continuous growth to decrease the crystallite size.[4]

Devadoss et al.[4] reported the changing trend of the crystal size in relation with the microstrain as follows:

- For the lower Zn content, lattice constant and the average crystallite size decrease due to the increase of microstrain which is responsible for the peak broadening, the reduction of the lattice imperfection and the existence of defects and vacancies.

- The increase of crystallite size at higher x is attributed to the decrease of strain which is caused by the predominant recrystallization process and the movement of Zn interstitial from the inside of crystallites to the grain.

The Zn^{2+} ions present both substitutionally and interstitially in $Cd_{1-x}Zn_xS$ thin film to result in the distortion of Cd–S lattice.[16]

The crystallite size decreases with the increase of Zn content as shown in Table 2.

The microstrain in $Cd_{1-x}Zn_xS$ thin film was calculated from the peak FWHMs by the following equation.[3,4]

$$\varepsilon = \beta\cos\theta/4. \qquad (5)$$

The strain increases with the increase of Zn content and we can find the inverse relationship between the decrease of crystallite size and the increase of strain.

Table 2. XRD measurement data and calculating structural parameters for $Cd_{1-x}Zn_xS$ thin film with different Zn content x.

Zn content, x	Peak position 2θ (°)	FWHM β (°)	Inter-planar distance d (Å)	Lattice parameter (Å) a	Lattice parameter (Å) c	Average particle size D (Å)	Micro strain ε (10^{-3})
0	26.8	3.74	3.326	3.841	6.652	21.846	15.867
0.1	26.9	3.78	3.313	3.826	6.626	21.62	16.033
0.2	27	3.83	3.301	3.812	6.602	21.342	16.242
0.3	27.2	3.87	3.278	3.785	6.556	21.13	16.405
0.4	27.5	3.91	3.242	3.744	6.484	20.927	16.564
0.5	27.7	3.93	3.219	3.717	6.438	20.83	16.641
0.6	27.9	3.92	3.197	5.537	—	20.892	16.592
0.7	28	3.9	3.186	5.518	—	21.004	16.504
0.8	28.1	3.86	3.175	5.499	—	21.226	16.331
0.9	28.1	3.82	3.175	5.499	—	21.448	16.162
1	28.2	3.77	3.164	5.48	—	21.737	15.947

Table 2 summarizes XRD measurement data and structural parameters calculated from those values for $Cd_{1-x}Zn_xS$ thin film with different Zn content x.

3.2. Morphological studies: SEM measurement

The observation of surface morphology was carried out for $Cd_{1-x}Zn_xS$ thin film with Zn content x from 0.2 to 0.8. Figure 3 shows the SEM images of $Cd_{1-x}Zn_xS$ samples with different Zn contents 0.2, 0.4, 0.6

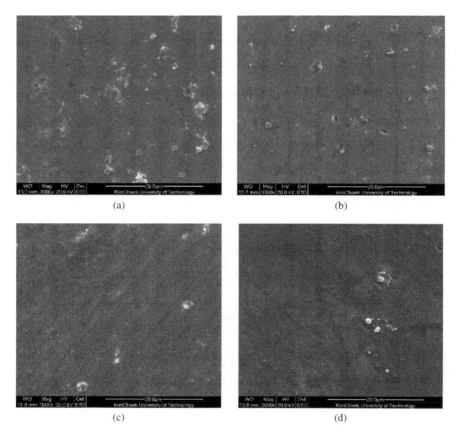

Fig. 3. SEM images of $Cd_{1-x}Zn_xS$ thin film with different Zn contents (a) $x = 0.2$, (b) $x = 0.4$, (c) $x = 0.6$ and (d) $x = 0.8$.

and 0.8. With the initial increase of Zn concentration, a certain size of granules was formed by the agglutination of Cd and Zn compounds around the nucleus due to the absorption of Cd^{2+} and Zn^{2+} ions, while the number of CdS nuclei decreases with the increase of Zn concentration, as a result, the size of granules further increase.

These granules with larger aggregates on the surface were formed by the addition of Zn during the reaction to change the surface morphology. During the formation process of CdS particles from $Cd(OH)_2$, some of these are stabilized to produce the nuclei while others are re-dissolved. The different sizes and non-uniform distribution of $Cd_{1-x}Zn_xS$ nanoparticles can be explained by that agglomerated grains of $Cd_{1-x}Zn_xS$ particles were grown starting with the colloidal nanoclusters. These clusters aggregate into larger secondary grains to minimize their surface energy; subsequently they collide and merge with each other to form multimers.[4]

On the other hand, the size of $Cd_{1-x}Zn_xS$ particle decreases and the mixture phase in thin film became stable for the samples with Zn content larger than 0.6, in agreement with XRD results, which shows the coexistence of the hexagonal phase and cubic phase. The increase of x decreases the number of large particles since the growth rate of small particles and formation of nuclei increase with the increase of Zn content, and thus it is likely that the particles with smaller size were produced from cubic $Cd_{1-x}Zn_xS$ or cubic ZnS.[6] This decreasing trend of particle size can be due to the smaller solubility of ZnS than CdS, which decreases as the concentration of Zn^{2+} in $Cd_{1-x}Zn_xS$ solid solution increases.[11]

As shown in Fig. 3, various sizes of aggregates (large white grain probably originated from CdS compound[7]) were found at the lower Zn concentration, and the size and number of aggregates decrease as Zn content increases. The surface roughness of $Cd_{1-x}Zn_xS$ thin film decreases with the increase of Zn concentration.

The existence of cadmium, zinc and sulfur was determined through the elemental analysis of $Cd_{1-x}Zn_xS$ thin films using EDX. Figure 4 shows the EDX spectra of $Cd_{1-x}Zn_xS$ thin films ($x = 0.2$–0.8). The peak corresponding to Zn spectra gradually increased and Cd spectra gradually decreased with the increased value of x.

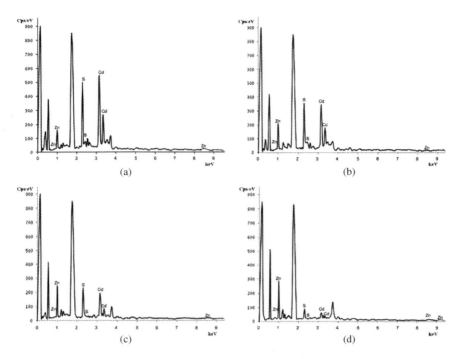

Fig. 4. EDX spectrum of Cd$_{1-x}$Zn$_x$S thin films for (a) $x = 0.2$, (b) $x = 0.4$, (c) $x = 0.6$ and (d) $x = 0.8$.

Table 3. Elemental composition of Cd$_{1-x}$Zn$_x$S film with the Zn content using EDX analysis.

Zn content in solution, x	Cd(%)	Zn(%)	S(%)	Zn composition in film	Zn/Cd	S/(Zn + Cd)
0.2	42.76	6.13	51.11	0.13	0.14	1.05
0.4	33.84	13.27	52.89	0.28	0.39	1.12
0.6	28.65	21.54	49.81	0.43	0.75	0.99
0.8	18.47	33.95	47.58	0.65	1.84	0.91

Table 3 summarizes the elemental fraction and compositional ratio of Cd$_{1-x}$Zn$_x$S film with the Zn content using EDX analysis. As shown in this table, the concentration of Zn in the Cd$_{1-x}$Zn$_x$S films increased gradually and the concentration of Cd decreased gradually.

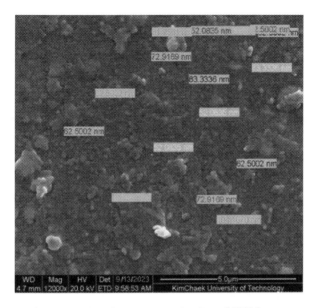

Fig. 5. The variation of average grain size of SEM measurement.

Figure 5 shows the SEM measurement image. As can be seen, the variation of average grain size was in ~100 nm range.

Figure 6 shows the cross-section SEM image of fabricated CdTe solar cell.

3.3. Optical properties

The optical transmittance of $Cd_{1-x}Zn_xS$ samples was measured in the wavelength range 300–800 nm, varying x from 0 to 1. Figure 7 shows the transmittance spectra of $Cd_{1-x}Zn_xS$ thin film with Zn content 0, 0.2, 0.4, 0.6, 0.8 and 1, respectively.

As shown in Fig. 7, the transmittance of $Cd_{1-x}Zn_xS$ thin film is higher than CdS film, while the average transmittance exceeded 85% in the visible region. The absorption edges are shifted towards the lower wavelength region as the Zn content increases, which indicates the increase of band gap.

The transmittance of $Cd_{1-x}Zn_xS$ film increases with the increase of x in the short wavelength region, which is caused by inter-band

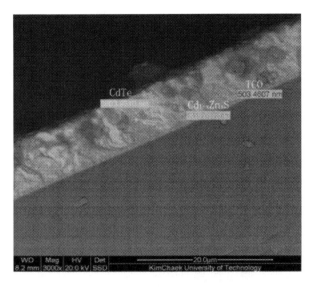

Fig. 6. The cross-section view of SEM image of fabricated CdTe solar cell.

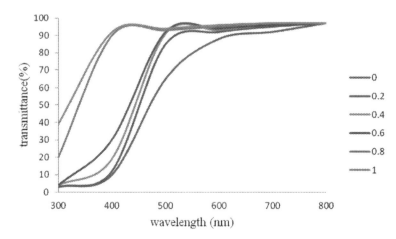

Fig. 7. Transmittance of $Cd_{1-x}Zn_xS$ thin films with the different Zn composition x.

transition from the valence to conduction band, whereas the film shows transparency in the long wavelength region.[4]

The increase of band gap implies that the particle size of $Cd_{1-x}Zn_xS$ film is in the nanometer range as the band gap gets wider at the smaller spatial dimension.[4] The band gap of nanomaterials, as the function of the composition and size, changes significantly at low crystal size. The increase of Zn content decreases the particle size and increases the band gap energy E_g. The initial shift of transmittance to the lower wavelength region was due to the substitution of Zn^{2+} in Cd–S lattice and the size effect. The enhancement of band gap energy at high Zn content is related to the size effect and the substitution of Zn^{2+} in Cd–S lattice, accompanying the improvement of crystallinity.

The quantum confinement effect begins to appear when the crystal size is less than the Bohr radius of the exciton (R_{BE}).[11] The quantum confinement is related to the relative increase in energy gap with the decrease of particle size, including the strong confinement regime (crystallite size $\geqq R_{BE}$) in which the band gap energy significantly increase with regard to the radius of crystallites and weak confinement one(crystallite size $\ll R_{BE}$) with smaller increase of E_g.[11] The bulk band gap energy of CdS and ZnS is 2.42 eV and 3.64 eV, respectively.

The energy gap of $Cd_{1-x}Zn_xS$ thin film was determined by plotting $(\alpha h\upsilon)^2$ following $h\upsilon$ values and then the straight line portions were extrapolated to the energy axis.[4] The energy gap of $Cd_{1-x}Zn_xS$ nanoparticles was found using the Tauc plot as shown in Fig. 8.

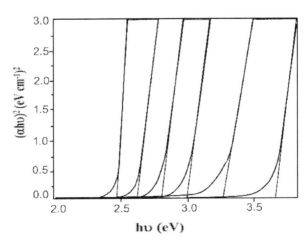

Fig. 8. The $(\alpha h\upsilon)^2$ following $h\upsilon$ values of $Cd_{1-x}Zn_xS$ nanoparticles for the energy gap values.

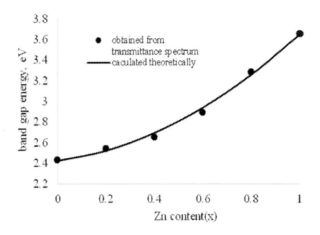

Fig. 9. Band gap energy of $Cd_{1-x}Zn_xS$ thin film.

The energy gap of $Cd_{1-x}Zn_xS$ thin films can be theoretically calculated in the range of $0 < x < 1$ using the following quadratic relation.[1,3,13]

$$E_g(x) = E_g(CdS) + [E_g(ZnS) - E_g(CdS) - b]x + bx^2 \qquad (6)$$

where $E_g(CdS) = 2.42$ eV, $E_g(ZnS) = 3.64$ eV, bowing parameter $b = 0.91$.

Substituting the energy gap data to a quadratic equation mentioned above yields

$$E_g(x) = 2.42 + 0.31x + 0.91x^2. \qquad (7)$$

Figure 9 shows the curve and values of band gap energy of $Cd_{1-x}Zn_xS$ thin films which were calculated from Eq. (6) and obtained from the transmittance spectrum.

Table 4 shows the E_g values of $Cd_{1-x}Zn_xS$ thin films with different Zn contents.

The increase of band gap at $0 < x < 0.6$ is due to the incorporation of Zn ion in the Cd site in the hexagonal CdS structure as the energy gap of hexagonal ZnS is larger than cubic ZnS,[6] while the change of band gap at $x > 0.6$ is related to the crystal phase

Table 4. Band gap energy of $Cd_{1-x}Zn_xS$ thin films with different Zn contents.

Zn content, x	0	0.2	0.4	0.6	0.8	1.0
Band gap energy obtained from transmittance spectrum, eV	2.43	2.54	2.65	2.89	3.28	3.65

Table 5. Resistivity of $Cd_{1-x}Zn_xS$ film with different Zn content.

Zn content, x	0	0.2	0.4	0.6	0.8	1.0
Resistivity, Ωcm	437	1536	3284	6283	8195	11,738

transition from the hexagonal/cubic mixture phase to cubic phase. del Valle reported that these change of E_g with Zn content is due to the increase in the position of the conduction band by the hybridization of Cd 5s5p level with the more negative Zn 4s4p level.[9]

The optical absorption in the long wavelength range (500 nm) is generally associated to the crystal defects such as grain boundary and dislocation, so low crystal defects density is needed to improve the transmittance of $Cd_{1-x}Zn_xS$ film.[6] Moreover, the sharp fall in the absorption edge is related to the defect of thin film, thus it is concluded that $Cd_{1-x}Zn_xS$ film with $x = 0.2$, 0.4 and 0.6 has good crystallinity.

3.4. *Electrical properties and solar cell performances*

The resistivity of $Cd_{1-x}Zn_xS$ thin film increases with the increase of x as shown in Table 5.

The increase of film resistivity with Zn content could be caused by the decrease of particle size, the increase of number and scattering of the grain boundaries.[3]

As several researchers reported, the electrical conduction is dominated in polycrystalline thin films by grain boundary scattering, oxygen chemisorption of chemisorbed oxygen on the grain boundary over 300°C of annealing temperature. In particular, oxygen absorbed in the grain boundary acts as a trapping center which is placed 0.9 eV below the conduction band.[12]

As the band gap energy of $Cd_{1-x}Zn_xS$ thin films increases with the increase of Zn content, the absorption edge shifts towards the lower wavelength range and more photons are absorbed by CdTe layer and then short circuit current is improved. As the open circuit voltage changes logarithmically for the short circuit current, the V_{oc} increases with the increase of x.[2] Based on the empirical equation of Green, the fill factor is dominated by the serial resistance R_S and the shunt resistance R_{SH}.[2] The R_s value decreases and the R_{SH} value increases with the increase of x. As the open circuit voltage and the fill factor aren't almost changed, the efficiency is dominated by the short-circuit current.

The enhancement of sheet resistance value may be connected with the change of the films' stoichiometry. It is known that the conductivity of II–VI compounds is greatly influenced by their stoichiometric composition.[12] Increase of sheet resistance as a function of the mixture ratio indicates that the incorporation and the substitution of Zn in the CdS lattice lead to increase in band gap.

Figure 10 shows J–V characteristics of the fabricated $Cd_{1-x}Zn_xS/CdTe$ solar cell with the change of Zn content.

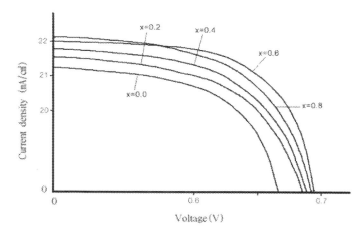

Fig. 10. J–V characteristics of fabricated $Cd_{1-x}Zn_xS/CdTe$ solar cells.

Table 6. Performance of CdTe solar cells with the $Cd_{1-x}Zn_xS$ window layers ($0 \leq x \leq 0.8$).

Window layer	Open circuit voltage, V_{oc} (mV)	Short circuit current, J_{sc} (mA/cm^2)	FF	Efficiency (%)
CdS	0.676	21.2	0.577	8.27
$Cd_{0.8}Zn_{0.2}S$	0.685	21.6	0.583	8.63
$Cd_{0.6}Zn_{0.4}S$	0.692	22.1	0.585	8.95
$Cd_{0.4}Zn_{0.6}S$	0.694	22	0.587	8.96
$Cd_{0.2}Zn_{0.8}S$	0.688	21.8	0.588	8.82

Table 6 shows the open circuit voltage, short circuit current, fill factor and efficiency of CdTe solar cells using the $Cd_{1-x}Zn_xS$ ($0 \leq x \leq 0.8$) as window layers.

As can be seen, better performances were obtained at $x = 0.4$, 0.6. This can be explained by the improvement of short circuit current caused by the increase in band gap energy and the enhancement of open circuit voltage due to the small number of pinholes and certain value of film resistivity.

4. Conclusion

In this work, $Cd_{1-x}Zn_xS$ thin films ($0 < x < 1$) were deposited by the CBD method and the influence of Zn content on the structural, morphological, optical and electrical properties of films was investigated by using the analysis instruments including XRD, SEM and UV-VIS spectroscopy.

The analysis results can be summarized as follows:

– XRD studies revealed that the crystal phase of $Cd_{1-x}Zn_xS$ thin film was changed from the hexagonal to hexagonal/cubic mixed phase at lower Zn contents, and then transformed to cubic ZnS phase as Zn content further increased. Moreover, it was found that the changing trend of the average crystal size and lattice constant of $Cd_{1-x}Zn_xS$ thin film was attributed to the shift of

peak position (2θ angle) towards the higher value, transformation of the crystal phase and the change in the intensity and FWHM of peaks with the increase of Zn content.
- Through the SEM measurements, we could observe the certain size of granules formed by the agglutination of Cd and Zn compounds around the nucleus due to the absorption of Cd^{2+} and Zn^{2+} ions with the initial increase of Zn concentration. On the other hand, for the Zn content larger than 0.6, $Cd_{1-x}Zn_xS$ particle size decreased and the mixture phase in thin film became stable.
- The transmittance of $Cd_{1-x}Zn_xS$ film increases with the increase of x in the short wavelength region by the inter-band transition, while the enhancement of band gap energy at high Zn content is related to the size effect and the substitution of Zn^{2+} in the Cd–S lattice.
- The increase of film resistivity with Zn content could be caused by the decrease of particle size, the increase of number and scattering of the grain boundaries.

From the above results, the optimal Zn content is $x = 0.5$–0.6. As the increase of Zn content, the absorption edge shifts towards the lower wavelength region in the transmittance as shown in Fig 7. So the use of $Cd_{1-x}Zn_xS$ as window layers in heterojunction solar cells instead of CdS leads to a decrease in window layer absorption losses. Also, the increase of resistivity in window layer leads to the decrease of the shunting effect which reduces the open circuit voltage and the fill factor. If the Zn content is too high, the heterojunction gets influenced due to the mismatching of lattice parameters and also of band gap. As a result, the efficiencies of solar cell decrease.

These research data would contribute to improving the performance of CdTe solar cell as effective window layers in photovoltaic applications.

Acknowledgment

The authors would like to thank the General Assay Centre, Kim Chaek University of Technology for carrying out the measurements of XRD, SEM and UV-Vis spectrophotometer of our samples.

References

1. G. Kartopu et al., Prog. Photovolt. Res. Appl. **22**, 18 (2014).
2. X.-B. Xu et al., Superlattices Microstruct. **109**, 463 (2017).
3. P. M. Parameshwari et al., Int. J. Nanotechnol. Appl. **11**(1), 45 (2017).
4. I. Devadoss et al., J. Mater. Sci. Mater. Electron. **25**, 3308 (2014).
5. S. V. Borse et al., J. Alloys Compd. **436**, 407 (2007).
6. T. Abza et al., Thin Solid Films **666**, 28 (2018).
7. R. N. Bhattacharya et al., Appl. Phys. Lett. **89**, 253503 (2006).
8. Z. Khéfacha et al., J. Cryst. Growth **260**, 400 (2004).
9. F. del Valle et al., Catal, Today **143**, 51 (2009).
10. I. Carreón-Moncada et al., Thin Solid Films **548**, 270 (2013).
11. K. Sreejith et al., Mater. Lett. **62**, 95 (2008).
12. Ng. Gaewdang et al., Mater. Lett. **59**, 3577 (2005).
13. S. Jana et al., Physica E **39**, 109 (2007).
14. A. J. Clayton et al., Mater. Chem. Phys. **192**, 244 (2017).
15. Y. C. Zhang et al., Mater. Lett. **61**, 4847 (2007).
16. L. Ma et al., J. Alloys Compd. **691**, 399 (2017).
17. Y. Dong et al., Mater. Sci. Semicond. Process. **19**, 78 (2014).
18. G. Laukaitis et al., Appl. Surf. Sci. **161**, 396 (2000).
19. P. Kumar et al., Opt. Mater. **27**, 261 (2004).
20. R. Zia et al., Optik **127**, 4502 (2016).
21. R. Hernández Castillo et al., Optik **148**, 95 (2017).
22. W. Xia et al., Sol. Energy Mater. Sol. Cells **94**, 2113 (2010).
23. G. Kartopu et al., J. Appl. Phys. **115**, 104505 (2014).
24. D. Joung et al., Nanotechnology **20**, 445204 (5pp) (2009).
25. Z. Zhou et al., ChemPhysChem **15**, 3125 (2014).
26. Z. Zhou et al., Phys. Status Solidi B **251**(3), 655 (2014).
27. W.-N. Wang et al., Appl. Catal. B Environ. **224**, 854 (2018).
28. S. A. Macías-Sánchez et al., Int. J. Hydrog. Energy **38**, 11799 (2013).
29. K. H. Chol et al., Opt. Mater. **112**, 2 (2021).

Index

ab-initio techniques, 94
absorption energy, 101
AC impedance, 4
activation energy, 2, 38
adsorption energy, 95
aerosol, 196
agglomeration, 149
annealing temperature, 78
anodization, 177
areal capacitance, 181
Arrhenius equation, 13
atomic force microscope (AFM), 185
atomic layer deposition, 111
atomic radius, 211

band gap, 176, 205
band gap energy, 219
biodegradability, 40
Bohr radius, 219
Bragg formula, 212
Brunauer–Emmett–Teller (BET) method, 113

capacity retention, 24, 203
carbon material, 64
carbon nanosheets, 67
cation exchange, 159
cell parameters, 4
cellulose acetate (CA), 26

cellulose-based separator, 23
cellulose separator, 39
charge–discharge profiles, 202
charge transfer resistor, 168
chemical bath deposition, 205
chronoamperometry, 121
clean energy, 1
close space sublimation, 208
collection efficiency, 115
constant phase element, 168
coordination, 209
corrosion resistance, 21, 28
Coulombic efficiency, 20, 202
counter electrode, 161
cryogenic-scanning electron microscopy (Cryo-SEM), 135
crystal defect, 221
crystalline water, 152
crystallinity, 219
crystallite size, 205
crystallographic structure, 98
Curie point, 58
current density, 157
cycle life, 24
cyclic voltammetry, 114
cyclic voltammograms, 87
cyclone collector, 196

2D material, 94
defect, 117, 212

dendrite growth, 20
density functional theory, 35
deposition, 149
dielectric capacitor, 54
dielectric constant, 27
differential scanning calorimetry (DSC), 141
diffraction peak, 201
diffusion coefficient, 114
dislocation, 221
double-layer capacitance, 169
dynamic light scattering, 135

electrical conductivity, 1, 33
electric double layer capacitance, 72
electric vehicles, 78
electrocatalytic behavior, 119
electrocatalytic oxygen evolution, 157
electrochemical active surface area (ECSA), 88, 169
electrochemical anodization, 188
electron transfer, 158
energy conversion, 111
energy density, 78
energy dispersive spectroscopy (EDS), 9
energy storage, 53
enthalpy, 146
environment-friendly, 64
etching, 187
eutectic phase, 137
eutectic salt, 137

Faradaic process, 78
field emission scanning electron microscope, 185
finite-size effect, 100
flower-like, 157
four-electron reaction, 158
FTO glass, 80, 207
functional group, 116
functional separator, 22

grain boundary, 221
grain boundary resistance, 11
grain resistance, 11
grain size, 5, 58
graphene, 26, 137
graphene oxide (GO), 26

half maximum (FWHM), 211
half-wave potential, 119
heat capacity, 138
heteroatom, 70
heteroatoms doping, 63
heterojunction, 206
hierarchical, 116
high magnification high-angle annular dark-filed scanning TEM (HAADF-STEM), 117
hollow flower structure, 69
hollow structure, 74
Hubbard interaction, 93
hybrid capacitor, 176
hybrid catalyst, 157
hydrogen production, 157
hydrothermal method, 28
hydrothermal reaction, 135
hysteresis loop, 59

impurity, 201
impurity phase, 55
initial charge capacity, 202
in situ characterization, 45
in situ synthesis, 158
intercalation, 79
intercalation/deintercalation, 85
interlayer, 19
intermediate membrane, 33
inter-planar distance, 201, 212
ion exchange, 41
ionic conductivity, 2
irreversible reaction, 202

kinetic current densities, 114
kinetic energy, 97
Kohn–Sham equation, 96
Koutecky-Levich equation, 114

latent heat storage, 136
lattice mismatch, 206
lattice parameter, 205
lattice spacing, 164
layered structure, 79
lead-free piezoelectric ceramics, 54
linear sweep voltammetry, 114
local current density, 24
long-term cycling performance, 202

Maxwell–Wagner effect, 27
mesoporous carbon, 138
metal air batteries, 157
metal–organic frameworks (MOFs), 24
microstrain, 212
Miller index, 212

MOF-808, 27
multiphase reaction, 202
multi-walled carbon nanotube, 137
MXene, 26, 93

Nafion, 114
nafion-based separator, 23
nanocarbon material, 137
nanoparticle, 118
nanosheet, 74
nanostructured thin-film electrode, 175
nanostructures, 78
nickel (Ni)-rich cathode, 194
nitrate combustion method, 3
non-Faradaic proces, 78
Nyquist plots, 168

onset potential, 119
open-circuit voltage (OCV), 94, 222
organic electrolytes, 20
orientation, 209
overpotential, 35, 157
oxygen evolution reaction (OER), 157
oxygen ion conductor, 2
oxygen-reduction reaction, 109

paraelectrics phase, 56
passivation layer, 20
phase change material, 136
phase separation, 147
phase transition enthalpy, 150
phase transition temperature, 149
pilot plant, 195

plane-wave approximation, 95
polarization, 54
pore size, 30
porosity, 27, 190
power density, 78
precipitation, 159
pseudocapacitor, 78
pyrolysis, 200

quadratic relation, 220
quantum dots, 109

Raman spectra, 57, 113
reaction kinetics, 74
recrystallization, 213
redox peaks, 201
redox reaction, 87
reference electrode, 161
relative density, 7
renewable energy, 64
Rietveld refinement, 56
rotating disk electrode, 161
roughness, 187, 205

Scherrer equation, 211
Schottky barrier, 158
sensible heat storage, 136
separator, 19
short circuit, 222
simulation, 45
single-atom, 122
single-atom catalyst, 109
single-crystal, 193
single-crystal nickel (Ni)-rich cathode material, 193
sintering temperature, 5, 53
solar energy storage, 136
solid–electrolyte interphase (SEI), 37

solid–liquid conversion, 146
solid–liquid phase transition, 138
solid oxide fuel cell (SOFC), 2
solid solution, 56
specific capacitance, 72
specific surface area, 137
spin coating, 25
spray drying, 196
spray pyrolysis, 193
strain, 213
stress–strain curves, 37
structural parameter, 100
substitution, 211
sulfonic acid groups, 41
supercapacitor, 63
supercooling, 137
surface functional group, 103
surface modification, 19
synergistic effect, 158

Tafel plot, 162
Tafel slope, 157, 162
tensile strength, 35
ternary compound, 205
theoretical calculation, 45
theoretical specific capacity, 94, 104
thermal conductivity, 138
thermal energy storage, 136
thermal evaporation, 77
thermochemical heat storage, 136
thin film, 79, 175
three-electrode system, 161
TiO_2 nanotube, 176, 188
Ton-scale preparation, 193
ton-scale production, 194
top-down approach, 188
transition metal compound, 158
transition metal oxide, 79

transmission electron microscopy, 68
transmittance, 217
transparency, 218
trapping center, 222
two-dimensional (2D) material, 24
two-step spray pyrolysis, 195

U-Hubbard term, 96
UiO-66, 27
UV-Vis absorption spectra, 113

vacancies, 212
van der Waals force, 100
van der Waals interaction, 100

Vegard's low, 212
visible region, 217
void cavities, 187

water splitting, 157
wavelength, 219
wettability, 35, 69

X-ray photoelectron spectroscopy, 69
XRD patterns, 4

Yang Shao-Horn principle, 158

Zn-Nafion separator, 36
zinc-ion batteries, 19

Printed in the United States
by Baker & Taylor Publisher Services